WRITING IS DES

WORDS AND THE USER EXPERI

Michael J. Metts
Andy Welfle

NEW YORK 2020

Writing Is Designing

Words and the User Experience

By Michael J. Metts and Andy Welfle

Rosenfeld Media LLC

125 Maiden Lane, Suite 209

New York, New York 10038

USA

On the Web: www.rosenfeldmedia.com

Please send errors to: errata@rosenfeldmedia.com

Publisher: Louis Rosenfeld

Managing Editor: Marta Justak

Illustrations: Nick Madden

Line Art: Michael Tanamachi

Interior Layout Tech: Danielle Foster

Cover Design: The Heads of State

Indexer: Marilyn Augst

Proofreader: Sue Boshers

ISBN: 1-933820-66-7

ISBN-13: 978-1-933820-66-8

LCCN: 2019950030

Printed and bound in the United States of America

MICHAEL

To Karina, Elena, and Elias

ANDY

To Katie and our large, fluffy sons

HOW TO USE THIS BOOK

Who Should Read This Book?

Writers. If you write the words your users read and interact with, this book will help you understand how to apply design techniques to your process. Whether you call yourself a writer, designer, content strategist, or something else, this book will help you write more effectively.

What's in This Book?

This book is full of ways to think about writing the words in an interface, along with strategic ideas you can apply to your work. You'll learn:

- How words shape design
- How to think about strategy and research
- How to write for clarity
- How to approach error messages and stress cases
- How to write for inclusivity and accessibility
- What the difference is between "voice" and "tone" and how to develop each for your product
- How best to collaborate with your team

What Comes with This Book?

This book's companion website (rosenfeldmedia.com/books/writing-is-designing/) contains a blog and additional content. The book's diagrams and other illustrations are available under a Creative Commons license (when possible) for you to download and include in your own presentations. You can find these on Flickr at www.flickr.com/photos/rosenfeldmedia/sets/.

FREQUENTLY ASKED QUESTIONS

What do you mean by "writing is designing?"

Just that. In many product teams, the words are an afterthought, and come after the "design," or the visual and experiential system. It shouldn't be like that: the writer should be creating words as the rest of the experience is developed. They should be iterative, validated with research, and highly collaborative. Writing *is* part of the design process, and writers are designers. That's the main thesis of this book (which you'll read in Chapter 1), and the point that we try to drive home in every chapter.

Is this book written only for writers?

No. Even if you only do this type of writing occasionally, you'll learn from this book. If you're a designer, product manager, developer, or anyone else who writes for your users, you'll benefit from it. This book will also help people who manage or collaborate with writers, since you'll get to see what goes into this type of writing, and how it fits into the product design and development process.

However, if writing is your main responsibility and you're looking for ways to collaborate with your team, you'll find those ideas in Chapter 8.

Will you teach me how to write error messages?

Yes indeed! We cover error messages and stress cases in Chapter 4. This isn't a how-to book, though—we talk about how to approach the work, how to think about it strategically, and how to set yourself up for success so you can jump in and do the writing.

What's the difference between voice and tone?

They're highly interrelated, but very different! "Voice" (Chapter 6) is the set of constant attributes in your writing that sets expectations, mood, and a relationship with your user. It's your product's personality. "Tone" (Chapter 7) shifts, depending on context: For example, you might write with a motivational tone if your users were new to your product, or with a supportive tone if they were frustrated. This book has strategies and approaches to developing those tone profiles and when to deploy them.

CONTENTS

FOREWORD

I wrote my first website copy in 2006. My client was a luxury condo development with a faintly ridiculous name, and my only source content was a print ad full of corny wine metaphors. I didn't care. I was 23 years old, and I was writing for a living. I even got benefits! So I sat down and clickety-clacked my way through: *Drink in the bold flavor of Bordeaux Heights.*

That's not real, but you get the idea. It was terrible.

Over time, I learned to be useful. To be clear. To make it obvious how things worked and where things were. I like to think I even got pretty good at it. But back then, I didn't have to think about how to make a complex onboarding flow feel intuitive, or which states we needed to consider in an app, or how to design content for a tiny screen. There were no iPhones—much less smartwatches or Fitbits or an app to control your thermostat.

A lot has changed. Interfaces now sit between us and all kinds of intimate moments and critical tasks. And each of those interfaces is full of words—words someone, somewhere has to write.

If you're that someone, this book will be your new best friend. Because designing interface content takes writing skill, sure. But it also takes curiosity about how things work, and compassion for the people they need to work for. And you'll find those things here, too.

Andy and Michael have a wealth of experience designing with words, and you'll see it in the coming pages. But what I love best about this book is that they know great interface writing doesn't come from lone geniuses with outsized egos. It comes from listening to as many perspectives as possible. In each chapter, they bring you fresh voices with essential knowledge on writing inclusive, accessible interfaces.

So dig in. Because this book is about more than words. It's about doing work that matters.

—Sara Wachter-Boettcher,
Author of *Technically Wrong*, *Design for Real Life*,
and *Content Everywhere*

INTRODUCTION

This book is for writers: People who want to use words to build better, more humane technology.

If that describes you, you're not alone. We've taught about this topic at events around the world. Each time, we meet amazing people who use their writing skills to help create digital products.

We interviewed more than 20 of those people with a variety of job titles for this book. The interviews were done individually, but we heard about some common challenges:

- Being left out of the process until it's too late to make an impact
- A perception that their work is easy and doesn't take much time
- Feeling less valuable than other team members

We've dealt with many of the same issues in our own careers. For us, the solution was not to get better at writing—it was to get better at user experience design.

Design is often perceived as visual, but a digital product relies on language. Designing a product involves writing the button labels, menu items, and error messages that users interact with, and even figuring out whether text is the right solution at all. When you write the words that appear in a piece of software, you design the experience someone has with it.

When writing is designing, the constraints are different. The goal is not to grab attention, but to help your users accomplish their tasks. The experience you're creating isn't permanent—it changes each time the team makes an update. And while you're likely working in a language that's familiar to you, your work could be translated into dozens or hundreds of other languages.

Here's what we think you'll need to be a successful designer of words:

- **An objective, strategic approach:** Take a step back from writing and focus on understanding. Research your users and find out what language they use. Learn about the people you work with and what they want to accomplish.

- **User-focused writing:** While mechanics and sentence structure are important, it's more important that your writing is clear, helpful, and appropriate for each situation. Don't think of your words as precious darlings, but as tools that can help your users, such as color, shapes, and interaction patterns.
- **Collaborative teamwork:** You'll need to help product teams understand your work and get involved in important decisions. As enthusiasm grows, you'll need ways to keep things consistent when lots of people are involved.

This is why we find it helpful to think of ourselves (and everyone who does this work) as designers. If you've never thought of yourself that way, we hope it's empowering and refreshing.

While this book is written specifically for writers, we think of that as a *role*, not a *job title*. Everyone benefits from learning how to design by writing. On many teams, designers, developers, engineers, product leaders, and others do the writing. Regardless of your title, this book will help you see how UX methods can be applied to what you write.

Whatever your background, we're glad you care about writing. We're excited that now more than ever, people are seeing how important this work is, but it didn't happen by accident. You'll hear from people throughout this book who have made it happen by influencing their teams and organizations.

You can make it happen, too. Stop writing clever copy. Start writing to design.

Our Story

We are a couple of those people we talked about earlier: *designers of words*. Writers on design teams. People who care about the language in the experience of digital products.

It's important to note that this isn't a tactical "how-to" guide. We believe that there's no one right way to do this kind of writing. Because of that, this book focuses on *how to think about doing the work* rather than *how to do the work*. Our goal is to give you ideas and concepts you can adapt to whatever you're working on.

We'll usually be speaking to you collectively, together, as one voice. But sometimes, if we have a personal story to share, or if we can add color to something relevant, we'll break out into our own separate identities. Look for pictures of us to read our personal stories and perspectives.

ANDY

Thinking about writing as a design practice was one of the most clarifying moments of my career. I've always been a writer at my core, but I've never been satisfied with just filling a space with words. I've always wanted to manipulate that space, too (in what I would call "design").

Back when I was in college, one semester, I was the editor of my university's student newspaper. One of my favorite things besides writing and editing was page layout. I really loved doing the following tasks:

- Deciding how to display articles
- Figuring out which articles should be more prominent than others
- Learning where and how to caption photos
- Deciding which quotes were important enough to call out in pull-quotes
- And on and on

Ever since then, I realized that I wanted to affect not just the words, but the system in which those words were used. I had no idea what UX writing was back then, or that it could even be somebody's full-time job to think about the words in software interfaces, but once I did, I realized what a perfect fit it was for my own interests.

Today, I work on a big, centralized product design team at Adobe, a prominent international software brand that makes tools for creativity. I lead a small team of content strategists and UX writers, and we're relatively new to this 30+-year-old company. Telling our colleagues that we "design with words" really helps them understand that we need to be involved at every step in the design process in order to really make a meaningful difference. ■

MICHAEL

I've always been interested in how words and visuals work together to tell a story. In fact, I studied art and communication in college, planning to become a photojournalist.

Unfortunately, I graduated at a time when no one was hiring photojournalists—but they *were* hiring people to work on their websites. I joined a UX team as a writer, and was surprised to find that many people didn't think of words as part of design. I make it my goal to help every team I work with see how words can affect the experience their users have.

Since then, I've had lots of different titles: UX Designer, UX Architect, and even Conversation Designer. I don't think you need a certain title to do this work well, but I do think many teams miss an opportunity to treat writing as a critical part of the design process.

My hope is that you're energized by the opportunity to use words to improve digital products. Whether you're a full-time writer, trying to become one, or just want to improve your writing skills while in another type of role, my experience has been that these skills benefit almost any product team. If you care about this work, you're the best person to help others understand how important it is. You can be the person who helps others see that writing is designing. ■

TOGETHER

We first met in October 2014, when Michael was teaching a workshop at Midwest UX that Andy was attending. We reconnected a year later at Confab, the biggest content strategy conference in the world and started chatting. Quickly, we realized that the thing we were missing from the conference was a track for UX writers. And then it dawned on us—why don't *we* create a workshop for those people?

Flash-forward to five years later, as we're writing this book. We've taught some version of this workshop six or seven times, and evolved it from a series of exercises on developing voice and tone for your product writing, to a comprehensive fundamentals workshop, covering much of what we talk about in this book.

While we live 2000 miles and two time zones away from each other and have very different career experiences, we still learn from each other all the time. We began to see that those differences are a good thing, for ourselves and for the people we teach. We hope that as you read this book, you'll see the value in your own experience and be empowered to share it. ■

CHAPTER 1

More Than Button Labels

How Words Shape Experiences

Two people stand in a conference room looking at printouts of mobile app screens. The office used to be a warehouse, but it's been renovated and turned into offices. The printouts are taped to the glass partition that separates their conference room from the hall, because tape doesn't work on exposed brick. It's a perfect stock photo opportunity.

"What does that button do?" asks one.

"It saves the user's data," says the other. "That's why it says 'Save.'"

"Does it save all their data, or just what we're looking at right here?"

"Oh. Just what we're looking at here."

"How will users know that? Should we tell them?"

Conversations like this happen all the time and in all kinds of places—not just repurposed warehouses. Teams who create software spend a lot of time talking about how people will use it.

That's where the word "user" comes from. People use buttons to take action, navigation to find where they need to go, or the dialog in a voice interface to figure out what their options are.

They also use words. Words help them figure out what that button will do, where that navigation will take them, or what that voice dialog means.

Start by Designing

How should those words be written? Most people have this question in their minds, but it's a tough place to start. Before you start writing, think about designing the experience you want your users to have. Here's how we think about these two activities:

> ***Writing** is about fitting words together.*
>
> ***Designing** is about solving problems for your users.*

To find the right words, writing *and* design need to team up in your brain and work together.

Think about the two people talking about the Save button in their mobile app at the beginning of the chapter. How should they know what to write?

- A **writing** mindset asks: How many words will fit here? How should I describe this action? What terms are we using elsewhere?
- A **design** mindset asks: What terms are our users familiar with? What happens next? What problem are we really trying to solve?

You can't have one without the other—and you need them both.

If the people you're working with don't understand that writing is designing, they'll be surprised when you suggest that changing how the experience works is the best way to improve it. Some problems can't be solved by writing, and learning to recognize when that situation occurs is just as important as learning to write a good button label.

Designing with words requires a broad range of skills, including many that don't involve arranging letters into sentences. Framing your work this way will make you more effective.

In our own work, we aim to design experiences that are usable, useful, and responsible. How does that apply to the words you write? Here are some questions you can ask yourself.

- **Usable:** Do the words help people use the interface? Are they clear? Do they help people accomplish what they set out to do? Are they accessible to all audiences?
- **Useful:** Do the words represent something people want to do? Do they give people control over the interface, product, or service? Does the experience add value to the user's life?
- **Responsible:** Could the words you're writing be misused? Are they true? Are they kind? Are they inclusive? Do they subvert language that people trust and understand to gain a business advantage?

To do this, you'll need to understand the product you're working on, along with the vision, constraints, interactions, visuals, and code behind it.[1] You'll need to spend time facilitating important conversations, conducting research, and aligning on a strategy.

Before you start writing, start designing.

---

1 In the software world, a "product" is what the team produces, not necessarily what they market or sell.

Usable Words

When a product is usable, it means that people can use it without coaching or help. You can find out if your software is usable through usability testing: giving users key tasks and then observing them to see if they're able to do what your product is designed to do easily.

But writing usable words goes deeper than that. For example, one of the best practices most people seem to know about interface writing is that you should *not* tell people to "click here." This advice is easy to remember, but the underlying concept is what's important. For example, it's easier for people to use a link rather than words when the link describes where the user will go. See Figure 1.1 for an example of how a link can make text more usable and clear. On the left, the words at the bottom of the list of tutorials tell the user to search for more and how to get to the search experience. The words at the bottom of the list on the right actually *take* the user to search. The words on the right are shorter and *much* more usable.

For the visually impaired people who use screen readers to narrate the words on a screen to them, this feature especially helps with usability.

Experts in the field of accessibility provide guidelines and best practices to support screen readers and more, but for the person using a screen reader on the "click here" link, what's the difference between accessibility and usability?

Sarah Richards, author of *Content Design*, gave a talk called "Accessibility Is Usability" in 2019.[2] She made the case that if the words you write for something aren't accessible to everyone, then you've made a design choice that prevents people from using that thing.

In her book, Richards pointed out that you can make writing more accessible and usable through plain language that people with a variety of reading levels can understand. This practice helps cognitively disabled users, those who have recently learned the language you're writing in, and even people who are stressed.

"It's not dumbing down," Richards said. "It's opening up."

2 Sarah Richards, "Accessibility Is Usability," filmed May 2019 at Confab, Minneapolis, MN, USA, video, https://www.confabevents.com/videos/accessibility-is-usability

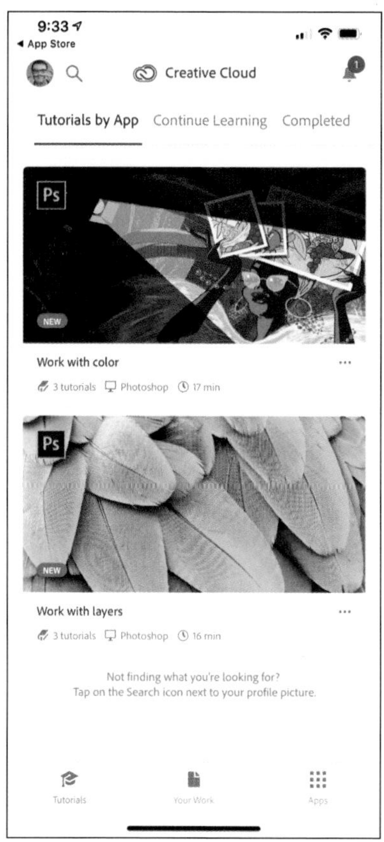

FIGURE 1.1
This before (left) and after (right) shot of an Adobe Creative Cloud mobile app search page shows some text at the end of a list of tutorials, prompting the user to use the search tool if their desired tutorial wasn't on that list.

Design experiences that are accessible to everyone include different literacy levels, cultural backgrounds, and disabilities. Usable writing works for all your users, no matter who they are.

Useful Words

For your words to be useful, you need to understand and honor the intent of your users. If you don't respect them, how can you expect them to keep giving you their time, money, and attention? Giving users control and prioritizing their needs is what makes writing useful.

Figure 1.2 shows checkboxes that appear as the user tries to pay for and reserve a hotel room. The first checkbox is required to complete the purchase. It requires that you sign up for the loyalty program, agree to the terms and conditions, sign up for marketing emails, and agree to the privacy clause all in a single checkbox. It also demands that users abandon their checkout flow to unsubscribe from email marketing.

The second checkbox frames opting out of emails as a negative option, so you may leave it unchecked because its logic is reversed from the first checkbox, which could lead some users to sign up for emails accidentally.

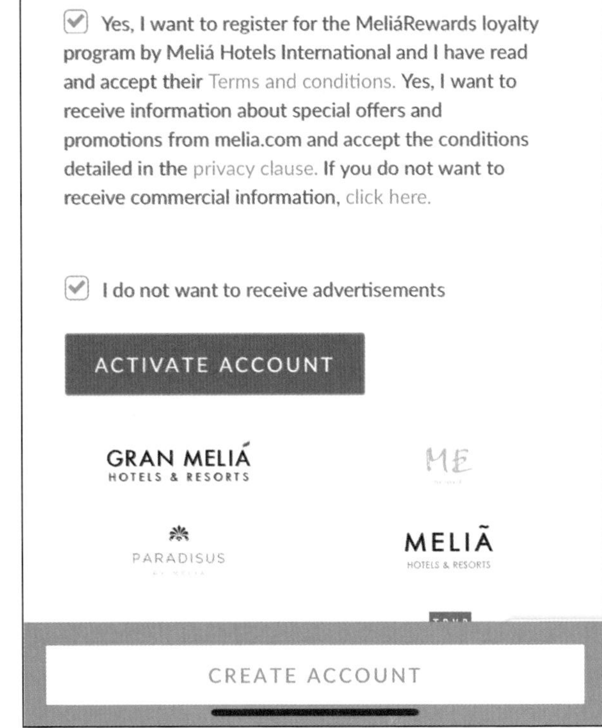

FIGURE 1.2
The tactics Meliá uses here to gain loyalty program members and email subscribers aren't a useful part of this room reservation system.

The team responsible for that reservation system didn't make creating a useful experience their priority. They used writing and design to force people into the loyalty program and email lists.

By contrast, the Pinterest Terms of Service show what happens when a team is able to build a vision for useful writing across a team that included designers, developers, and lawyers.

Figure 1.3 shows a portion of their Terms of Service, which includes a summary of each section in simple terms, to help users understand what they're agreeing to.

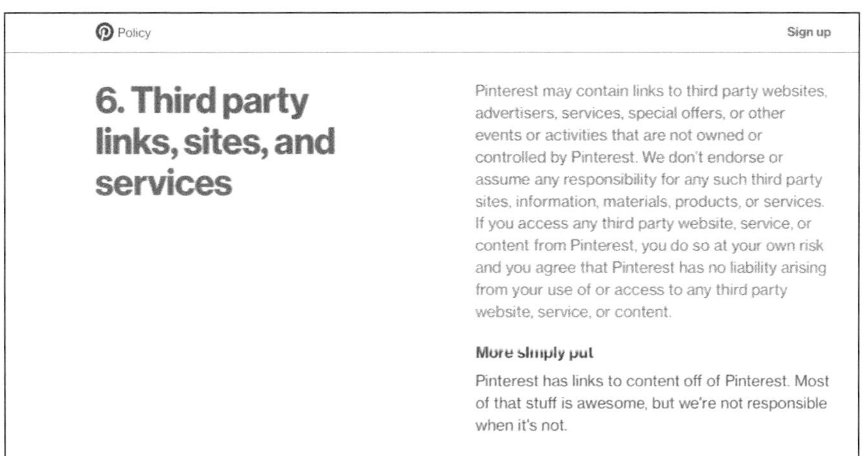

FIGURE 1.3
Pinterest's Terms of Service uses words to design an experience that's far more useful than most of the legal agreements users are forced into.

Useful writing focuses on what people want from your product or service and asks how you can balance that with business goals, rather than focusing purely on what the business can get out of it.

Responsible Words

Words should be used for good. As a writer *and* as a designer, you're responsible for what you put into the world and your words have power. To write responsibly, you have to consider a wide variety of scenarios.

Irresponsible writing weaponizes language to cause harm to your users. One popular example is the practice of "confirm-shaming." This situation occurs when an interface asks the user for something and then forces them to say something negative about themselves to decline. In Figure 1.4, a news app call theSkimm forces users to say they prefer to be miserable in the morning (see the last line) to close a form that asks for their email address. (Have you noticed how companies really desperately want email addresses?)

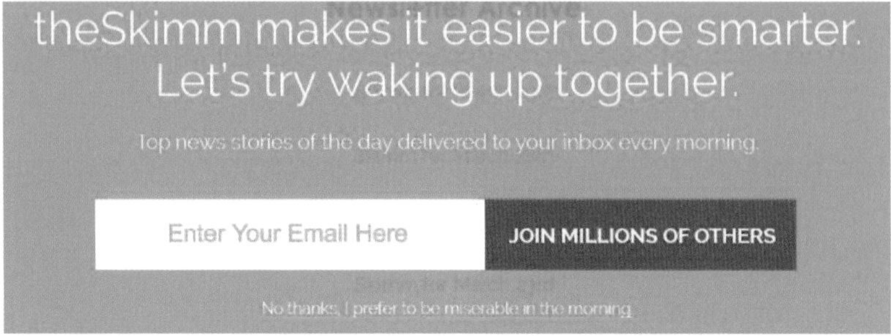

FIGURE 1.4
No one prefers to be miserable in the morning, but theSkimm forces people to say they do. This example was found on **confirmshaming.tumblr.com**, which collects examples of this.

But it goes deeper than not being a jerk. Often, words can cause harm in less obvious ways.

Figure 1.5 shows a LinkedIn conversation between two people. The first person is offering help to someone who was recently laid off. The second person responds with gratitude and says they're going to take some time to process what happened.

LinkedIn's algorithm suggested that the first person respond with a pre-built message like "Congratulations!" or "Sounds good!", which wouldn't have been appropriate. This feature is likely designed to help people save time, but in this case there's something far more important at stake than saving 30 seconds on typing a message. An accidental tap could have made this interaction hurtful and insensitive.

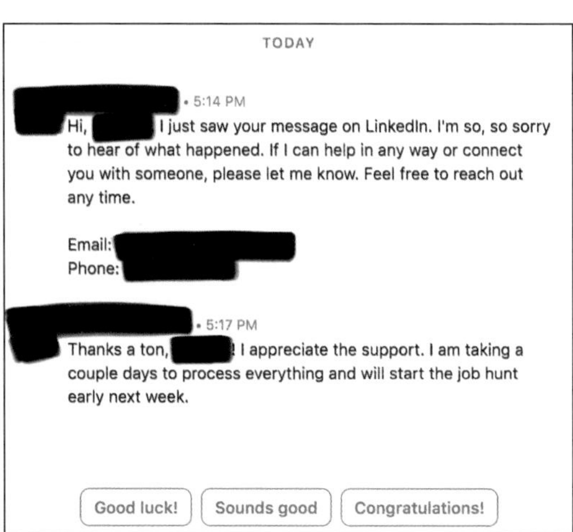

FIGURE 1.5
"Good luck!" is remarkably insensitive when someone has recently lost their job. "Congratulations!" doesn't make sense at all.

These pre-built messages probably aren't being used the way the writer intended, but that doesn't let them off the hook. As a writer, it's your responsibility to consider not just how your writing will be used, but also how it could be misused, whether it's by an algorithm, others in your company, or malicious people outside it.

How Words Build Experiences

What does it mean to design something by writing? It means that your words are building someone's experience.

Here's an example that has nothing to do with writing or software: Think about leaving your house to buy some apples from the grocery store. If you're physically able to walk, and the store isn't too far away, and your neighborhood was designed to include a sidewalk, it's easy to take a stroll to get your produce.

But what if you live in a suburban area, with few sidewalks? These neighborhoods are often designed to include sidewalks in a subdivision, but not along the road. So going outside the subdivision involves a choice between walking in the ditch or walking in the same area as cars that are driving at a high speed. In this case, most people choose to drive to pick up their groceries.

What if you can't walk? If you get around in a wheelchair but the intersections in your neighborhood weren't designed for wheelchair access by including curb cuts, you may have to drive or rely on others to run your errands (see Figure 1.6).

FIGURE 1.6
A curb cut is designed to enable wheelchair users to move easily from the street level to a sidewalk. It also typically includes tactile paving that helps visually impaired pedestrians know when they are about to enter or exit a street.

All of these combinations of roads and sidewalks have been designed, but who designed them? Who is responsible for being so accommodating to cars or choosing not to accommodate wheelchairs? Was it the person who drew up the plans for the roads? The government that approved their construction? The developers of the subdivision? The estimator at the construction company? The workers who built them? The answer is that all those people had a hand in designing them. When you make decisions that affect the experience someone else has, you're designing.

Nicole Fenton, coauthor of *Nicely Said*, describes her work this way in her article *Words as Material:*[3]

> I work on digital products and physical goods, so I'm deeply involved in the design process. But I also want to call out early that my process is the design process. I don't write fiction or short stories; I use language to solve problems—whether that's behind the scenes or in the product itself. I use words as material.

Words build digital experiences, and this book is all about creating good experiences for the people who use software on their computers, phones, watches, and other devices. More and more often, people use that software for personal, everyday tasks: paying bills, sending emails, or requesting a rideshare like Uber or Lyft. You're designing the interfaces that let them do that.

You can't create these experiences without words, and every word included in those experiences shares the user's understanding, feeling, and outcome. That's why this kind of writing is design.

The Words Are Everywhere

Right now, pull out your phone and open up one of your favorite apps. Take a moment to tap through it and take note of all the words you see.

That app you use every day relies on words. Yet, as you can see in Figures 1.7 and 1.8, if you take away the words, how much interface is left?

3 Nicole Fenton, "Words as Material," Nicole Fenton, March 12, 2015, https://www.nicolefenton.com/words-as-material

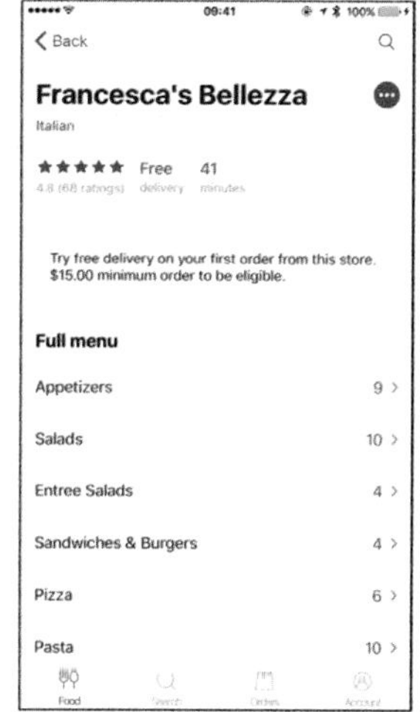

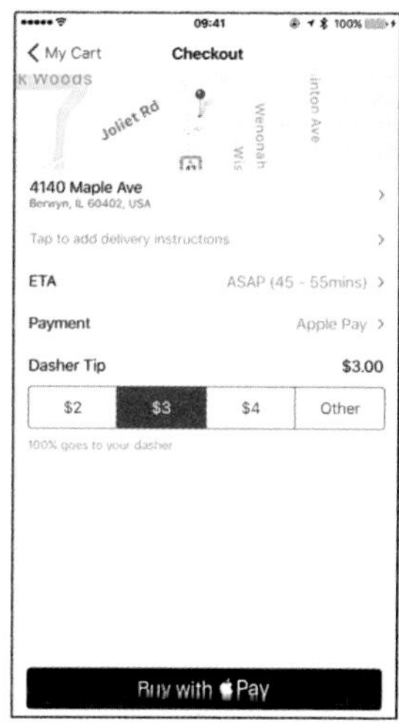

FIGURE 1.7

Screens from the DoorDash mobile app with words.

FIGURE 1.8

Screens from the DoorDash mobile app without words. Designer Mig Reyes did this to popular websites in a 2015 blog post, illustrating how much interfaces depend on writing.[4]

4 Mig Reyes, "Reminder: Design Is Still About Words," *Signal v. Noise*, January 17, 2013, https://signalvnoise.com/posts/3404-reminder-design-is-still-about-words

Think, for a moment, about that food delivery app—the one that would cease to function without words.

There are countless companies that will deliver food to where you live. These companies accomplish this through software products that have to be designed, developed, and then used by customers.

The people who use these products call them "apps," and that almost makes it sound like someone's weekend project, but this is serious business. Grubhub, for example—one of the biggest food delivery companies in the U.S.—pulled in over a billion dollars of revenue in 2018. Figures 1.9, 1.10, and 1.11 show a few of the different ways a product like this is used.

FIGURE 1.9
Customers often order from a mobile app (exactly which app depends on their operating system), but could also use their internet-connected watch, voice-activated assistant, smart TV, or anything with a web browser.

FIGURE 1.10
Restaurants receive orders through a different app (usually running on a tablet or laptop next to the cashier), and the people who deliver the orders use the mobile version of the restaurant app to see customer addresses and update the delivery status.

FIGURE 1.11
Offices may offer group ordering for their employees, so admins need a place to choose the restaurants, employees need a place to order before the cut-off, and the accounting team needs a place to view the records so they can audit payroll deductions.

It's a complex ecosystem of software, business, and people working together to get that cheeseburger to your door quickly and easily. There's a lot to write, too. Here are some of the components in a food ordering app that rely on words:

- App store listings (for each app store)
- Release notes for new versions
- Onboarding information (orientation instructions for new users)
- Login screens and forms
- Account recovery mechanisms
- Account areas and settings
- Payment screens
- Button labels and interface elements
- SMS notifications
- Push notifications
- Email notifications
- Confirmation emails
- Account recovery emails
- Email validation emails (emails about emails)
- Re-engagement emails
- Help content
- Terms and conditions
- Privacy policy
- Contact forms
- Contact form confirmation screens and emails

That's not even an exhaustive list—each company handles it differently. The words that appear in these areas are part of the experience of ordering food now, and someone has to write them.

The Need for Writing

It's not just about the apps on your phone, either. Any web app or website has interactive elements driven by words.

Every button has a label. Every form has error states. Every sign-up process has instructions. The words are everywhere, and it's a mistake to treat them as an afterthought—something that can be filled in later.

In fact, when it comes to certain types of interfaces, words are all there is. For natural language technologies like smart speakers and chatbots, there are rarely—if ever—visual elements involved in the design process (see Figure 1.12).

FIGURE 1.12

You don't need any fancy prototyping software to create a design for this thing—just a text editor.

Teams in this situation rarely need someone with design experience in the traditional sense. They need writers with design experience.

Writing the words in an interface first, before any kind of visual design, helps a team understand what they're working on and gives them something to respond to.

Writers won't be replacing visual designers anytime soon (or even at all), but on teams where the writer and designer are two different roles, both should be aware that they're serving the same users and working toward the same goals. It's not a question of one or the other. One doesn't come earlier or later. It's a collaboration that creates the user experience.

Katie Lower is very familiar with this type of collaboration. She has been working with words on various digital design teams for more than 15 years, and she's found it empowering when her employers recognize her work as design. "I think it gives you confidence—you just show up differently," she said. "I felt like if someone was calling me a designer, it leveled the playing field for me."

Lower didn't set out to design things, but she wanted to make a bigger impact on the product she was working on and the team she was working with. On a project early in her career, a usability specialist shared research findings with her that dictated what words customers were looking for at a certain point in the experience. She became curious about not just what she was supposed to write, but why she was writing it in the first place.

"I know we have these different specialties in design, but that experience, it all felt too chopped up and disconnected. The bigger picture got lost," she said. "It made me feel like there was something more to this work than being directed and told 'We need words in this space, put words in this space.'"

She pursued a Master's degree in Library and Information Science, which gave her an understanding of information architecture—the importance of content structure and how systems retrieve information—all of which are relevant to digital design.

"I wanted to be more than just the writer, and decided at that point in my life another degree was one way to get on that path," she said. "For me, it was a way to gain some experience and confidence."

Because Lower was focused on writing, the biggest challenge she faced was being brought into a project too late to have an impact. She would often ask questions to understand the decisions that had been made, and sought out teams that would welcome those questions as part of design.

"I feel like I always need the full context of what I'm solving for, so it's best for my work when I'm able to be in environments where I can get it," she said. "If you're joining a project at the very end and there's low tolerance for questions, it's a sign your role as writer hasn't been well positioned or isn't well understood."

Don't Apologize for Yourself

True story: A group of UX team members sat around a table, getting ready to participate in a design-thinking workshop. It's the kind of session where the team generates ideas for how to improve a user's experience with a product.

All but one person in this room had the word "designer" in their job title at some point: the person who didn't have the title was a writer.

The group opened with an exercise where they were each supposed to create a rough sketch of a solution that would improve a piece of software. One by one, each person presented their idea and explained how it would help the user. When the group got to the writer, he began with an apology. "Sorry," he said. "I'm not a designer, but here's my idea."

It's a small thing, but it's important. This person clearly felt self-conscious, even though he was hired to be on a UX team at a large company. He was in this meeting because his ideas were good. In fact, his design sketches were just as good as everyone else's, but at some point in his career, someone made him feel like his identity as a writer made his skills inferior to his teammates.

The point isn't that you should immediately start conversations with your manager about changing your job title or that you should insist on people referring to you as a designer. Job titles are fluid, and most often, they depend on internal company politics.

The "writer" is whoever is doing the writing. The writer could be someone who specializes in the craft: a UX writer, content strategist, content designer, or one of many other variants. The writer could also be someone who works primarily in another discipline: a designer, developer, product manager, or a UX researcher.

Titles don't matter when you're trying to do the writing. What matters is that your team delivers words that meet the needs of your users throughout their experience. Writing is part of delivery. To create good software, you need words in it.

Some of the work of a writer involves giving other people guidelines or coaching so they can do their own writing. That's important work, but it still requires that you think through the problem from the perspective of a writer who is designing the experience—someone who is using words to give meaning to each user's experience.

Describing writing as design isn't a power play. It's a reality of any UX writer's day-to-day work and an important way to approach your job. Like any other designer, you should be doing a lot more than writing. The best way to spend your time may be by researching user needs, prototyping solutions, testing your ideas, creating a strategy, or building rationale for your decisions. You might even be leading the product team and driving alignment around the direction of the product development.

Treating writing this way is new for a lot of people (and maybe you're one of those people). That's okay. It just means that your first task is to embrace your identity as a designer and help your team see it as well. Explaining your work and why it's important isn't an exercise in navel gazing or stroking your own ego. It's part of helping everyone involved understand how words fit into the experience you're creating.

Build Better Places

As you design with words, you're creating digital places where people spend their time. It's a big responsibility. One person who has spent a great deal of time thinking about, working with, and writing about using language this way is Jorge Arango. He's an information architect and the author of two books on the subject.

Arango believes that one way to learn how to use words more effectively is to learn another language. "The reason I advise that is that it forces upon you, at a very deep level, an understanding that language is contingent on historical factors we take for granted," he said. "Language is so important to us, and we acquire it so early on, that we can lose sight of the fact that it is a construct, and one that is evolving."

However, Arango believes that writers have valuable skills to offer the technology industry—especially when it comes to creating names and labels for digital products. "I suspect that most people come to these decisions with vocabularies that are not as broad as the job demands," he said.

As an example, Arango described how something like a News Feed (used by Facebook and others) brings certain user expectations. "News is the feedback mechanism of our society; we vote based on the things we learn in the news," he said. "When we take a concept like that and we subvert it for commercial use, that's something that should give you pause."

This is the greatest responsibility you have when you use writing to design experiences. You're not simply coming up with labels for buttons and navigation—you're changing how your users think.

"Persuasion is a powerful thing," he said. "If you are the person who controls the form of the environment by defining its boundaries through language, the persuasion will happen without me even knowing it's happening."

Writing the user experience may be difficult at times. It's a skill that's often underestimated and undervalued. However, it's exactly what the world needs.

Finding What's Right for You

Just like a designer designs the user experience, writers write it. There are many ideas in this book that will help you do the work, but it's even more important that they help you *think* about the work. Your needs are unique, so don't try to find one right way to do things. There isn't one.

Instead, find what's right for your users and your team. Adapt these ideas to your own work, and then expand on them and develop them. That's how you go from writing to designing.

CHAPTER 2

Strategy and Research

Beyond Best Practices

"'Purchase' is better right? 'Buy' just sounds so cheap. This is a sophisticated product." A group of smart, capable people are having a debate about word choice.

Maybe this has happened at your company: product owners, designers, marketers, engineers, and maybe even you, the writer, getting hung up on all the different words you could use to ask your users to give you money.

During one of these meetings, a leader remembers that they have someone in the room who is paid to come up with words. "What do you think?" they ask. "You're the writer. Do you know what other companies say in this situation? What's right? 'Buy' or 'Purchase'?"

This kind of conversation can be frustrating. People are always looking for the right way to write something. They want to know the best practice.

That's understandable. It's easier to defend a decision by saying, "Don't worry folks, I wrote it this way because it's a best practice in the industry."

But who benefits from a best practice?

Your users don't. They want to get things done. They couldn't care less if Google and Amazon do it.

Businesses think they will benefit from a best practice, but it usually just gives them an easy way out of a problem (or a meeting). Those people having the "Buy" vs. "Purchase" debate are probably frustrated. They've got lots of stuff to do, and they're spending their time discussing word choice.

So what's the right answer in this scenario? Surely one option is better than another.

The answer is that it's not about being right or wrong. It's about what's right for your situation, what works for your users, and what your team is trying to accomplish. Strategy and research bring all of that together.

Before you can figure out whether "Buy" or "Purchase," or even some other word ("Get?") is the right choice, you have to do some work to understand what's appropriate.

Align on Your Strategy

Even if you're not the one creating or communicating strategy, you need to understand it and point your team to it. Working without a strategic focus feels like trying to put out a house fire with a garden hose. Even if you somehow succeed, it will take a long time, and you'll feel defeated along the way.

You don't need anyone's permission to be strategic. You can start right now.

Strategy starts with aligning people. Alignment is a *businessy* word that just means everyone is on the same page. It's really important, but it's surprisingly hard.

Alignment happens all the time outside of work—it's ingrained in society. Can you imagine driving up to an intersection if all the other drivers weren't aligned to the idea that they should obey the traffic signal? It could mean injury or death.

It's rare that the stakes are that high in software development (although it becomes more of a concern every day as things like cars and medical devices become digital), but a lack of alignment definitely wastes time and money. Left unchecked, it could mean that you don't meet user needs and the product fails. All because we don't take the time to make sure we understand each other.

Here's how Kristina Halvorson, author of *Content Strategy for the Web*, (and someone whose work has been formative to our careers) talks about strategies:

> They're directional statements that tell teams, "Here's where you're going to focus your efforts for a defined time period." In other words, strategy tells you what to do . . . and, by default, what not to do.[1]

Just talking about strategy can benefit your team. Often, each person has a strategy in their head, but they don't share it with anyone. Multiply that by the number of people on your team, and you can see why alignment is important. Teams are central to Halvorson's definition because strategy only works when everyone is aligned.

---

1 Kristina Halvorson, "What Is Strategy (and Why Should You Care)?" *Brain Traffic* (blog), September 21, 2017, https://www.braintraffic.com/blog/what-is-strategy-and-why-should-you-care

Halvorson knows the value of strategy for this type of work. Her book is widely recognized as required reading for anyone working with web content.

When the idea of content strategy first became popular, it was born out of the need to control the chaos of content. Every company, from the mightiest global corporation to the lowliest subsidiary, was getting a website, and they were filling it with stuff. Often, it was stuff their users didn't need or want. Content strategy serves as a compass for organizations in that position, and greatly improves the web by focusing on user needs.

But content strategy is about more than cutting out the unnecessary or ineffective bits from your website. It's about giving a large group of people a common framework for their decisions. It's organizational rationale.

Digital products and services are no different. They're filled with words of their own— interface content. Just because the words are part of an interface doesn't mean you shouldn't be strategic about them.

Start with your team. Everyone—from engineering to testing to product leadership to design—needs to know and understand the strategy.

Depending on your organization, this could include the following people:

- **Executive leaders:** These people understand their business. If they're executives of a medium-sized or large business, it's how they got promoted again and again. They are the people with the power to pull your funding or change your focus. These decision-makers greatly benefit from understanding your strategy—and you benefit from their support.
- **Marketing:** Brand ambassadors can become brand police when they feel threatened. Help them see how your strategy portrays the product in a better light and fits into what they're already doing.
- **Legal:** Chances are, when you ask a lawyer if you can do or say something, they'll say no. That's because if they say no, there's no risk! Find ways to get them excited about product direction, like holding workshops or inviting them to research share-outs where they can see why your strategy matters to the business.

IN PRACTICE

Strategy Statements for Products

A great way to get teams aligned is to create a strategy statement with them. Statements like these are valuable to the whole team, but they're especially important for writers. As a writer, you may be asked to describe a product's features or increase conversions during an onboarding flow. Strategy statements give you direction that the team believes in, and help you find some of the answers you need to write about the product effectively.

You're in the perfect position to help product teams create these strategy statements, because they're made of words. *And what are you good at?*

Spoiler: Words.

Content strategy consultant (and author of this book's foreword) Sara Wachter-Boettcher created a Mad Lib exercise to help teams collaborate on a statement, and Meghan Casey wrote about it in her book, *The Content Strategy Toolkit*. Casey's adaptation focuses on four things:

- Business goal
- Content product
- Audience
- User need

This makes sense when content is the product (like on a website), but you can adapt Casey's method to focus on the value that a product brings to its users. Use these starting points:

- User type
- User need
- Value provided to the user
- Business benefit

Let's say you're working on a product that helps users learn how to play the drums. There's a mobile app that shows the lessons, a website where you manage your account and billing, and even a smartwatch app that tracks activity during practice sessions. It's a startup, so it'll be named Drum.ly or Drum.io or something. If you're facilitating a strategy statement for the entire product, your Mad Lib might look like Figure 2.1.

continues

IN PRACTICE (continued)

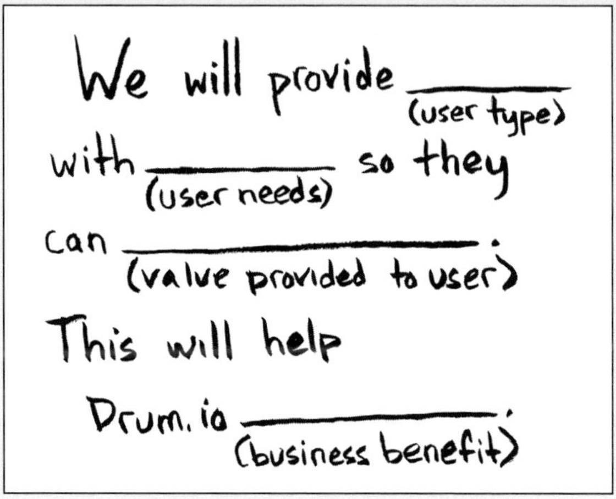

FIGURE 2.1
A strategy statement Mad Lib written on a white board for a team to fill out.

Next, ask the team to fill in the blanks in small groups. Give them a few guidelines to make sure that the activity is effective.

1. **Ask each group to designate a facilitator.** Their job is to make sure that everyone in the group agrees to the decisions being made.
2. **Each group should think of 1–2 specific things per blank.** The benefit of this exercise is that it forces teams to articulate what's most important to them and gives leaders a chance to hear that from their teams.
3. **Ask each team to share their statement with the rest of the group.** This will spark valuable conversation and help everyone see the patterns that emerge.

After doing this in small groups, you can take what they made and create a strategy statement that reflects the desires of the whole team.

Here's how the statement for Drum.io's strategy could turn out:

> We will provide new drummers with a simple way to learn drumming basics so they can play along with other musicians. This will help Drum.io gain adoption and build revenue.

This statement reveals a lot. If your goal is to get your users to the point where they can play with other musicians, that affects the flow you create and the way your team prioritizes features.

Knowing the key business metrics of adoption and revenue helps the team see how the product and its marketing should work together.

If you apply it to the "buy" vs. "purchase" scenario we started with, the team now understands that they're trying to reach a wide audience within the general public. This could include a wide variety of education levels, people who speak English as a second language, and people with cognitive disabilities. To reach these users, keep your language as accessible as possible. In fact, the Readability Guidelines project (yes, that's a real thing, a collaborative effort led by Content Design London) recommend "buy" in this scenario for that reason, and provide a lot of research to back up their choice.[2]

Keep this statement in front of the team. Refer to it in meetings and conversations. Evaluate your decisions through it and encourage others to do the same.

It's important to note that this doesn't only have to be used for big, product-level strategy. You can use it at the feature level, or even for an initiative that your team is trying (like a new batch of push notifications). Anytime you find that your team needs a shared strategy, this collaborative method can help.

What does a good strategy look like? Here are some questions to ask yourself:

- **Is it actionable?** Does the team know what to do? Do they know how it applies to them?
- **Is it relevant?** Does it fit into the larger goals of your product or organization? Does it make sense in light of what other teams are doing?
- **Is it user-focused?** Will your users benefit? Is it based on research?
- **Is it verifiable?** Will reasonable people agree when it's done?

If you answer no or even *maybe* to any of these questions, it's worth taking a second look at your strategy.

Why Your Strategy May Not Be Effective

The whole point of spending time on strategy is that it should impact the work you're doing. If it's not, then you might want to revisit it.

Maybe it's not opinionated enough. For example, a strategy like, "We will make a good app with good features that will help our users and help our company meet its goals" is a nice sentiment, but it's not really saying anything useful. Be sure to lead your team in a direction where they're talking about specific features and taking a stance on how to make sure those features are useful for the user and good for the company.

---

2 "Readability Guidelines," Content Design London, last modified August 6, 2019, https://readabilityguidelines.myxwiki.org/

This isn't an exhaustive list, and a lot of it depends on your organization. Whether your company has three employees or 30,000, strategy should be bringing them together.

The point is, you're unlikely to get anything done by declaring your strategy and then defending it. The fastest and most effective way to do the work is to make sure that everyone believes in it just as much as you do.

In fact, strategy shouldn't be about you at all. That's usually where you can get into trouble. If people feel like you're pushing your own agenda, they'll have very little motivation to follow you.

Nobody gets away with loudly declaring their very smart opinion and then watching everyone line up to follow them. Consultants, if you think you can get away with it, take it from us, who have worked on both sides: Nobody opens up that PowerPoint deck of recommendations after you send it over with the final invoice.

Find Answers Through Research

Aligning on a strategy is a great starting point, but finding the right words also requires researching your users and testing your ideas with them. While your first impulse may be to write what you're asked to write, learning is usually the right place to start.

Imagine for a moment that instead of writing for software interfaces, you're a travel writer. You show up to work on your first day excited to explore new places and go on adventures.

You get your assignment: A feature story on Indonesia. However, you're told that when you're doing the research for your article, you can't go there and aren't even allowed to talk to anyone who lives there. Your boss tells you that Brad from Sales went there once, so you should just ask him about what it's like.

As silly as this sounds, this scenario is surprisingly common on product teams. People are hired to create a good experience for their users and then told that there's no time to talk to, learn about, or spend time with them.

That doesn't make sense. In fact, one study by User Interface Engineering showed that when each team member spent two hours in contact with their users every six weeks, the quality of their work increased dramatically.[3]

Research is never perfect, but it should also never be optional. It's the best way to understand your users and their needs.

Escaping Opinions

People who build software talk about their users all the time. They say what their users want, what they expect, and how long they're willing to wait. They explain why the team should do something by starting sentences with "Well, if I was the user . . . ."

This is great if they actually know something about the user, but usually, they're just conveying their opinion. Don't let opinion dictate the words you're writing. Instead, move past the biases you and your team have and rely on insights and data about your users to make decisions.

You won't just be researching words. What you write will be far more effective when you do the following:

- Learn more about the people who use your product and their environment
- Understand whether users are able to use the experience your team has designed
- Get an idea of how people perceive the product and react to it

You can't create a good experience for your users if you don't know much about them. You need to do your research. Whether you work with a research team or educate yourself, research is too important to skip.

3 Jared M. Spool, "Fast Path to a Great UX—Increased Exposure Hours," UIE, March 30, 2011, https://articles.uie.com/user_exposure_hours/

FINDING THE RIGHT PROBLEMS

Once, I was working on a product that asked customers for information about things they own. These were high-value items like cameras, bikes, and computers.

My team created a slick flow designed to gather all the required information from our users as quickly as possible. It was beautiful, had nice animations, and followed accessibility standards.

The problem is that it didn't work.

By testing the design with our users, we discovered that many of them stopped and gave up because they weren't sure how to provide what the form was asking for.

Figure 2.2 shows the portion of the form that asked for photos of the items before testing. There were labels, but they weren't providing enough information for users to move forward.

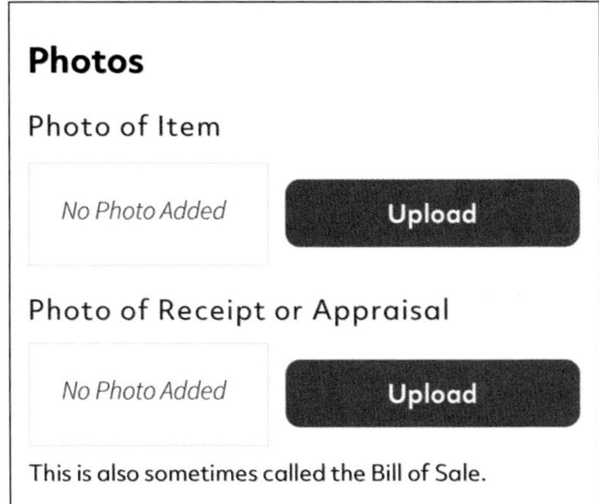

FIGURE 2.2
These labels were descriptive, but testing showed us they didn't provide enough information.

In order to help users complete these tasks and upload the right photos, we needed to be more descriptive, explaining what we needed and why.

After testing, we adopted a solution that had a lot more explanation (Figure 2.3).

When we first started talking about these form fields, I raised the concern that we weren't explaining enough. I felt that we were relying on too much jargon to guide users through this form. While the team didn't agree at the time, testing made it incredibly obvious.

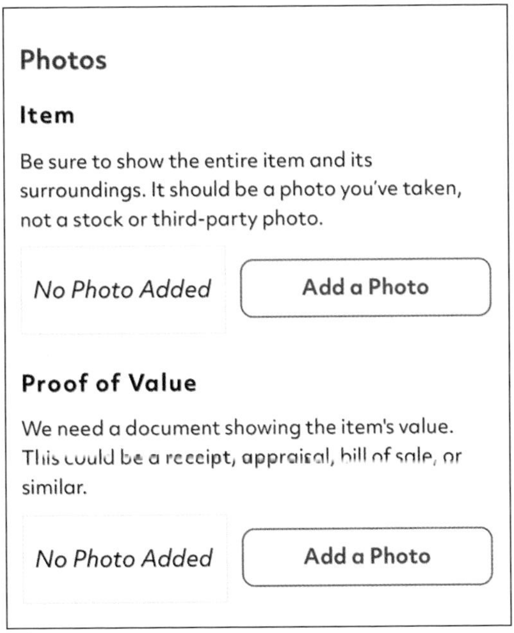

Photos

Item

Be sure to show the entire item and its surroundings. It should be a photo you've taken, not a stock or third-party photo.

No Photo Added | Add a Photo

Proof of Value

We need a document showing the item's value. This could be a receipt, appraisal, bill of sale, or similar.

No Photo Added | Add a Photo

FIGURE 2.3
The update included a lot more wording, but helped users move throw the flow more easily, because they understood better what the app was looking for.

If you're having trouble helping your team adopt a user-centered approach to writing, testing can be an incredible way to help them see why things need to change. ■

Research Questions

The best starting point for any research effort, before you even talk to users, is to write down your questions. What do you want to learn?

Here's an example: If you're working on a product that helps professionals create diagrams and drawings, here are some questions that could help guide your research.

- What types of work situations require people to create visual artifacts?
- How do people create diagrams and drawings for work?
- What are the demands our users face through their jobs?

Whether broad or narrow, research questions will help you and your team understand what you want to learn; then you can plan activities that will help you answer those questions.

User Interviews

Interviewing users helps you understand how they talk about the concepts you're working with. Matching their language increases usability.

Whether you're interviewing users on your own or working with a researcher, ask questions that will help you uncover how people refer to the things your product refers to and take special note of the words they use.

In the example of the diagramming tool, a question like "Tell me about something visual you created for work" will help you understand your user's vocabulary without leading them (see Figure 2.4). In fact, this term leaves out the "drawing and diagramming" language that's part of how your team describes its product, leaving them free to describe that visual artifact however they'd like.

Interviews also help you learn more about what it's like to be your user and how they behave in the situations you're designing for. In the process, you'll discover their language and how to write for them.

FIGURE 2.4
Keeping your internal language out of interview questions will gives users the freedom to describe things in their own words. What you think of as a diagram, they may think of as a chart.

NOTE ASKING THE RIGHT QUESTIONS

With interviewing, it can be easy to use poor techniques and gather bad data. Using leading questions or cutting off participants before they have time to respond leads to insights that are unreliable.

As important as research is, it's even more important to learn how to do it correctly. It's a specialized field, so if you're going to dive in and start interviewing people, be sure to learn how to do it effectively and responsibly.

If you want to learn more about interviewing, there are a lot of great books on the topic. One of our favorites is *Interviewing Users* by Steve Portigal.[4]

GOOD INSIGHTS TAKE TIME

In working on a redesigned share menu for graphic design software, a researcher, a designer, and I interviewed several graphic design users ranging from creative professionals to amateur designers. We were trying to understand how and when they used words like "post," "publish," "share," "export," and so on.

We didn't say that we were listening for their vocabulary, of course. That might have led to them become more careful with their words and skewed the results. We just asked them about their workflow while they talked us through working in the app.

The interviews were time consuming, but they led to really great insights. For example, for most users, "post" was a much more casual and less final sharing action than "publish," which felt more formal and finished.

I updated my writing accordingly. ■

Contextual Inquiry

Another technique to learn about your users is contextual inquiry. It involves going to where your users are and observing their behavior in their own environment. You may ask questions, but the focus is on learning and observing what happens in your users' world. By focusing on behavior and understanding the environment, you'll be able to more clearly see the problems that your product can solve and how it could be used.

4 Steve Portigal, *Interviewing Users: How to Uncover Compelling Insights* (New York: Rosenfeld Media, 2013).

This is important for the product in general, but it's also important from the perspective of working with words. When you immerse yourself in someone else's world, you begin to understand their needs so much better.

TOUGH CONDITIONS, CHALLENGING DESIGN

I once led contextual inquiry with a team working on software designed for construction workers on commercial job sites. I visited over 10 construction sites in the United States and Canada, and gained a new appreciation for the conditions these folks were working in. They were frequently required to wear gloves, would always need protective gear, and often dealt with intense heat that would make devices slippery from sweat and fog up their safety glasses.

In fact, to be on these sites, I had to do my research with the same protective gear, taking notes on a clipboard. Rather than a traditional interview guide, I created double-sided worksheets that helped me remember the details I wanted to capture each time we observed one of the participants (Figure 2.5).

FIGURE 2.5

The protective gear and interview guides I used for contextual inquiry on a commercial construction site. Confidential information has been blurred.

Doing this helped the team understand that whatever we built for these users needed to prioritize speed and comprehension above all else. It also helped us shift our strategy from focusing on the workers themselves to building features that managers could use to help their whole teams be more efficient. After all, because of the conditions, the workers would rarely have the time and ability to use the app we were planning to build.

We never would have had these insights if I hadn't taken the time to be in their environment. As we created a strategy for our product, we referenced the strenuous conditions again and again—internalizing our users' needs. ■

Contextual inquiry isn't strictly about the words, but it will still help you know how to write them. You will start to understand the world that your users are living in and how you can use words to help them.

Usability Testing

If you ever feel your team starting to disagree about a detail of implementation and the meeting starts slipping into a fog of uncertainty, chime in with "Why don't we test this and see how it works for our users?"

Usability tests consist of listing key tasks for the user to complete, observing them as they use the software to do those things, and describing their experience along the way. In these cases, you're testing every aspect of the experience at once—the visuals, interactions, words, and, in some cases, even the performance of the code.

As a writer, be sure that the tasks you're testing include scenarios that rely on the language you've written. Include confirmation screens, error messages, and any situations where users need to understand the words.

Here are some ways you can be involved in usability testing:

- **Observe or run the sessions.** You can learn a lot from watching users interact with an experience. Learning how to plan and run a test will benefit you even more.
- **Take note of language.** Listen to how users describe what they're doing and what they're looking for. You can incorporate those terms into your writing later.
- **Identify information gaps.** Look for moments where users are confused or are looking for information they can't find. These are opportunities for design updates you can work through with your team.

IN PRACTICE

Content Testing

One of the best ways to test how your users perceive and understand what you've written is to remove it from the interface completely and test it by itself.

Content designers on the gov.uk team work to make sure that the words on British government websites are user centered. In one of their testing methods, they print out the writing, asking users to highlight the sections that make them more confident in green and sections that make them less confident in red.[5]

We've adapted this technique to test the words and messages in product interfaces. Usually, we add some narrative to help participants understand the scenario.

You can also change what you're asking participants to evaluate. One variation is to ask users to circle what's helpful and underline what's unhelpful. This is especially good for evaluating the tone of your writing, since the results will often point to words and phrases that don't add any value to the interaction.

Asking follow-up questions is a critical part of any testing effort, and it's often where you'll find the most value. Here are some things you can ask about:

- **Motivation:** Asking why participants describe things as helpful or unhelpful will help uncover needs you may not have been aware of, as shown in Figure 2.6. You may find that important details are missing, or that particular terms are confusing for your users.
- **Expectation:** Ask users what they expect to happen next. You'll learn a lot about how they interpreted what you wrote, and whether it's clear enough to be usable.

5 Pete Gale, "A Simple Technique for Evaluating Content," *User Research in Government* (blog), September 2, 2014, https://userresearch.blog.gov.uk/2014/09/02/a-simple-technique-for-evaluating-content/

Because writing is such a core part of the experience, doing usability testing alongside your team will almost always be beneficial, but there are other testing methods that can help you with writing in particular.

Depending on what you want to test, there are a variety of testing approaches you can use to find the appropriate words.

- **Perception:** Ask users to rate the message verbally based on your desired outcome. For example, if you've established the voice of your product as professional, you could ask them to rate the message from 1 to 7, where 1 is casual and 7 is professional.

FIGURE 2.6
Asking follow-up questions after testing is often the most important part of a test. In this example, the user reveals that when dealing with billing, an exact date is important to them.

The Cost of Not Doing Research

We've heard lots of objections to doing research, including the following:

- **No time** (We need to build this thing—not spend more time spinning!)
- **No money** (Who's going to fund it?)

- **No access to users** (We are very protective of our customers—you can't talk to them.)
- **No need** (We already know what our users need.)

These are terrible reasons, and you shouldn't let any of them stop you from learning about your users and what they need. In almost every case, skipping research will cost your team more time and money than doing it.

Building a user-centered experience doesn't happen through intuition. Knowing about your users and their behavior will help you find what's appropriate for them more quickly, making your writing process more efficient and effective.

Beyond that, research will begin to establish your user as the primary stakeholder. This is an important transformation for any team. At the end of the day, insight from users should be prioritized over anyone's opinion—including your own.

Finding What's Right for You

Often, your first impulse will be to write something, but we're starting this book with strategy, research, and testing because these things help you find the right words for your situation. These are the things that root your work in the user experience, increasing its effectiveness.

We've only scratched the surface of research methods in this chapter, but it can feel overwhelming. There are many research methods we didn't even touch on, including quantitative methods like A/B testing and surveys. There are also methods rooted in information architecture, like card sorting and tree testing, that will help you figure out how users organize and categorize things.

Strategy and research are what make this type of writing unique. They help you know what to write and whom it's for, providing a solid foundation for your writing and design decisions.

CHAPTER 3

Creating Clarity

Know What You're Designing

One thing many writers have a strong opinion about is the serial (or Oxford) comma. If you're unfamiliar, it's the comma that comes before the "and" in a list, as in "this book is about writing, designing, and the user experience."

Every major style guide on writing takes a firm stance. (*The Associated Press Stylebook*, for example, is against using it, but the *The Chicago Manual of Style* is for it.) It's common to see writers declare their personal stance in their Twitter profile.

"Without it," proponents cry, "There will be chaos! No one will know to what we're referring in lists!" Then they point to an example of an author dedicating their book to "my parents, Beyoncé and God."

On the other side, the anti-serial comma faction pipes up. "But that's why we have context clues! We all know it's implausible for someone to believe God is one of their parents! Plus, we could just reorder that list to 'God, Beyoncé and my parents'! That comma is redundant and therefore unnecessary! We must be concise!"

As with all strong opinions about subjective things, the answer lies somewhere in the middle. Context matters. If you're writing a legal document, where precision is key, a comma might mean (and has meant) the difference between winning or losing a lawsuit.

If you're writing a newspaper article, with a finite amount of space and a thin column of text, it's probably not needed. That's why newspapers follow the *The Associated Press Stylebook*, which focuses on brevity and concision.

Whether or not you're using the serial comma doesn't matter so much as how much clarity it adds to your message. And to understand that, you need to understand the *context* of your message.

Know What You're Writing

While your team may not care about words as much as the serial comma crowd, it's important to help everyone understand the role they play.

In the past, ad copywriters worked with designers and account executives to synthesize the product's value proposition or feature messaging into copy. And no copy was more important than that clever tagline—it was the hook that drew readers' eyes to the ad (as you can see in Figure 3.1).

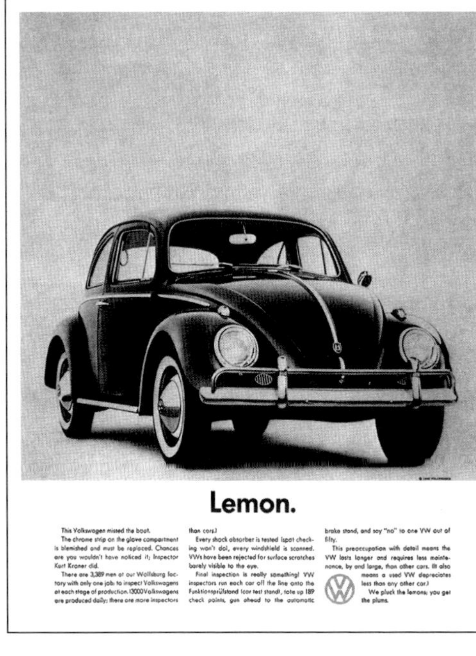

FIGURE 3.1
A 1960 ad campaign for the Volkswagen Beetle, where copy plays the largest role in hooking and captivating the reader.

But when you're writing for a digital interface, it's different. Are you trying to educate a user? Are you trying to intrigue them into exploring more features? Charm them into upgrading to a premium tier? Or are you trying to make a complicated feature simpler?

Bring the Clarity

Writing is designing, so your primary goal should be to bring clarity to your users. Clear writing helps users understand what's happening with the product and what it means for them.

To bring clarity to your users, you have to start by bringing clarity to your team.

If your head is filled with questions about what your product or project is, you may feel embarrassed. Shouldn't I know what I'm writing for? Look at all these other people—don't they know what they're making?

Well, it may surprise you, but no, there's a good chance they don't. Sure, they know what features they're building or what interface they're simplifying, but often, especially with enterprise software, even the project lead hasn't seen the full breadth of the site or application.

But words have a way of facilitating clarity. Here are some examples of things you may need to write for a product, along with questions you can ask to understand them more clearly:

- **Introductory screens and onboarding:** Who are our users? Why should they care about this product? How does it make their lives better? What bits could be confusing?
- **Billing systems:** When does the first payment happen? When do payments renew? What methods are accepted?
- **Notifications:** What's the goal of this notification? What is our team being measured on? How does this help the user?

These are questions that everyone may have in their heads, but writers are sometimes the first to ask them. It's hard to write about something without understanding it. While you need clarity if you want to do your job well, your team and your users will benefit from it, too.

Design teams use software like Sketch or XD to show what they're designing—and those tools are great—but it's easy to get caught up in the details. There are sliders for rounding the corners of buttons and color pickers for precisely choosing which shade of blue they are, but there's no selector for figuring out what you're working on or why it matters. It's an open world—a blank slate. It's a job for words.

A lack of clarity isn't just frustrating for members of the team, it's also expensive. When everyone has a different understanding of the problem you're trying to solve, they may ship a feature that isn't needed or put up a web page that has to be revised a few days later.

So before you start writing button labels, or working on voice and tone guidelines, use the world's most underrated and effective design tool: a text editor.[1]

Start by writing. Write without constraints or voice or tone.

Detail what the interface should say. If you do that, you'll understand better how it works. Words are the key to unlocking meaning, and because of that, it's perhaps the most powerful way to align a group of people with different perspectives, agendas, and preferences to a project.

1 You can use a word processor like Microsoft Word or Google Docs, but we recommend something that supports plain text like Windows' Notepad or Apple's TextEdit app. The important thing is not to worry about formatting—just write!

Understand Implications

If you design a digital product without focusing on the words, it's easy to avoid thinking about the message that it sends to its users. When you take responsibility for the words in an interface, you are the last line of defense against manipulative practices, misinformation, and jargon. Thinking about the language intentionally means that you're also thinking about its ethical implications—from how it could exclude people to the way it's used to shape and influence behavior.

In her book *Technically Wrong: Sexist Apps, Biased Algorithms, and Other Threats of Toxic Tech*, Sara Wachter-Boettcher writes about a smart scale that emails you when you step on it. By default, the message it sends when your weight is higher than it was before is disappointed, but encouraging: "You've gained X pounds. Better luck next time!"

To someone who is trying to lose weight, the message is fine. Harmless, even. But what about someone with anorexia, or a child, even, who's trying to gain weight? It's an eye-rolling message at best and harmful at worst.

Somebody had to write those emails. If they were thinking systemically about the words they were writing—not what the words should be, but also how they fit into the larger system of words the user sees—they might identify the cases where users aren't experiencing what the company thinks they're experiencing. Of course, even if the writer understood these implications, the organization they work for may not be structured to allow them to flag potential concerns and change course once the project is underway. Large, siloed organizations tend to be bureaucratic and process driven, and aren't particularly adept at changing strategies downstream. It could be that this writer wasn't empowered to raise this concern.

That's why it's important to collaborate with your team on the words as soon as possible, while a project's scope is still new and malleable. It's much easier to change wording—or even the product's message—when it's written on a white board or a Word doc than when it's coded into an interface.

Recognition and Recall

Being clear is challenging on its own, but the real challenge is to do that over and over again, throughout the experience you're working on.

The Nielsen Norman Group, an organization that conducts user experience research, has a list of ten important usability heuristics for user interface design. (A heuristic is, broadly, a technique that allows someone to teach themselves something.) One heuristic that's key for writers to understand is "recognition over recall"—in other words, make plain the object, action, and options a user has, so they don't have to remember it. One of the best ways—perhaps the best way—is to do that with words (as in a citation).

The study digs deep into memory retrieval, particularly the concepts of recall and recognition:

- **Recall**, a process your brain uses to retrieve information from your memory, is how you remember things like usernames, passwords, web addresses, and other open-ended prompts.
- **Recognition**, on the other hand, draws from many more overt clues to allow you to make a decision. An example might be a button with a clear call-to-action, for example, or a list of menu items. Each of these elements gives you the cues you need to recognize an option, make a decision, and ultimately accomplish your goal in an interface.

For example, if you label an action "Delete" in one area and "Erase" in another, the words may mean the same thing, but now the user will have to pause whatever they are doing and think about it. Choosing one option (along with rationale to back up your decision) and using it consistently will help your users recognize the action when they see it, while being assisted by whatever they're able to recall from the last time they used the action.

Visual layout and presentation is important for recognition, but the words you use are especially important. The clearer and simpler they are, the more you give the user cues to borrow from their own understanding and experience to move forward.

Lightening the Cognitive Load

Making things simple for your users can be a big challenge. One person who is up for that challenge is John Saito, a product designer at Dropbox, a San Francisco–based software company that sells cloud storage services and creative tools. Like many writers working in user experience, his path wasn't a direct one: he worked as a localization writer, a help documentation writer, a UX writer, and more. Early on, as a UX writer, he read *Don't Make Me Think: A Common Sense Approach to Web Usability* by Steve Krug.

"It completely transformed the way I think about words," Saito said. "The basic idea is that you need to try to make the user think and read as little as possible. That's the only way to have a fighting chance at people even reading your content and digesting it. It's always in the back of my mind.

This leads him to work as hard as possible to reduce the number of words and reduce the number of choices that are presented to a user at once.

"If, as the writer, I have to think about the words for more than two seconds, I know it's not good copy."

Saito said that when he was in college, he studied cognitive science—how people think about the world. Soon he found himself in a class with George Lakoff, a prominent cognitive linguist, and was digging deep into the idea of metaphor in language.

"It turns out the way in which we understand the world is almost entirely through metaphor," Saito said. "If you closely study the words we use, you can trace it back to metaphor."

When it comes to design, Saito says, everything is a metaphor.

"The gestures we use—clicking, tapping, or swiping, they're all metaphors of something we'd do in the real world. And the icons we use—floppy disks for saving, and the cutting and the pasting. All metaphors."

Words are the same. You can use metaphors in terms, like "inbox" and "timeline," or in actions like "copy" and "paste," as Saito mentioned. Some products are built entirely on metaphors, like Photoshop. It's full of tools related to photo editing and compositing, with its roots in darkroom photography and tabletop editing.

NOTE **JOHN SAITO ON METAPHORS**

Metaphors help people make sense of your product. Saito says there are a couple things to keep in mind if you're developing a metaphor:

- Does this metaphor actually describe what the user is trying to do?
- Don't make me think: Is it common enough for a broad understanding, or is it too localized for my user base to understand it?
- Is it internally consistent, or does it conflict with other terms and actions in my product?

There's nothing wrong with these metaphors, as long as your audience can still borrow understanding from the original concept. Some take on a life of their own. (For example, we still say that we'll "dial" a telephone number, even though the majority of people alive today have never used rotary phones with a literal dial.) Some metaphors lose their meaning over time. (Many users today have never seen a floppy disk and don't understand how it means "save.") And some products evolve past their original metaphor and become internally inconsistent. (For example, Photoshop has a "magic heal brush" that uses intelligent algorithms to erase elements from a photo, extending well beyond the capabilities of an analog photo shop.)

HOW METAPHORS INFLUENCE BEHAVIOR

While a metaphor often helps users, it can sometimes make things more challenging. Voice assistants and chatbots rely on the metaphor of conversation to work, which means that you may have to design extra features just to accommodate how users think about the interface.

Figure 3.2 shows one of the first conversational interfaces I designed. It greeted users and then asked what they needed, which is exactly how you'd expect a human providing chat support to act. Users would greet the bot back to be polite, but the bot was expecting them to describe their issue. This meant that we needed to design a new opening message that didn't prompt this type of response and helped our users faster.

In this case, the metaphor of conversation changed user behavior, and we had to make sure it didn't go too far past what the technology was capable of. ■

9:41
Back
Support Chat
Done
Chatbot at 10:32pm
Hi, I'm a chat bot. How can I help you?
Jose at 10:36pm
hi
Chatbot at 10:38pm
Sorry, I didn't catch that. What do you need help with?

FIGURE 3.2
This chatbot was expecting users to describe their issue, but the conversational metaphor prompted them to greet it instead.

FINDING THE BALANCE BETWEEN PRECISION AND CONCISION

When you're building a complicated product, often you're faced with a complicated message and enormous constraints. One time, I was working on a search results page for a social networking app. The directive I was given was that this particular section of the search results showed profiles that matched the following parameters:

Here are profiles matching your query, showing:

- *Friends*
- *Friends-of-friends*
- *Those who have written posts that friends or friends-of-friends have liked or commented on*

And the constraint I faced was that I had one line on a smartphone screen to show this message. (Oh, and I needed to leave space for at least 30% of the screen's width for when it was translated into longer languages.)

Impossible. I immediately realized there was no way I was going to be able to convey all of these nuances in four or five short words. After several iterations and frustrating writing sessions, I returned with this:

People connected to you

This was the absolute best I felt I could do under the circumstances. I wasn't capturing a lot of nuances, but if we painted with broad brushstrokes, this made sense. I turned it in. It was concise, and it fit the space allowed.

I didn't hear about it again until an engineer reached out to me. She was starting to put this functionality together, and she had major objections with that heading.

"This tells the user nothing about the functionality of this module! We need to be more precise," she argued.

I countered: "This module is inherently too complicated, and there's no way to capture that precise functionality in a short string of text! We either need to be concise, or we need to simplify the functionality."

We were at an impasse. This was an extreme example of when a message was too complicated to properly express in copy—and to stay within our constraints, much of the message was lost. (See Figure 3.3 for an example of a spectrum of messaging to explore the balance between concision and precision.)

FIGURE 3.3
We often have to find a balance between being precise and concise, which can sometimes be in conflict with each other. This example shows how you can explore those characteristics with a message to find what's right for you.

That's a situation where we couldn't solve it with words. We had to go back to the drawing board to revisit, fundamentally, what the module was trying to be.

This module never shipped, but it did eventually evolve into separate modules—with simpler, more direct functionality. ■

Clarity Is in the Eye of the Beholder

It sounds like a no-brainer, but it's important to use language that your audience understands. Your users should be at the center of your writing process, and that's especially true when it comes to the language you use in an interface. It should make sense to them, and ideally, include the words they already use. In addition to the research methods in Chapter 2, "Strategy and Research," here are a few ways to do that:

- **Online discovery:** If you find your users online, you'll also find a wealth of information about how they communicate. Maybe the people who use your software hang out in certain Facebook groups. If you're working on a mobile app, check the reviews to see how your users describe it and what they use it for.[2]
- **Search analytics:** If your product or site has a search feature, analyzing what people search for doesn't just help you understand the words they use, it can also help you identify areas for improvement within your product. Often, a few of the top search terms reflect the majority of what people are searching for.[3]
- **Communication analysis:** If you work for a large company, you'll often find information about your users' language in places like call-center logs or support tickets. Build relationships with those teams and see if they can share their data. By breaking that data down, you'll begin to see the most common patterns emerge, and you'll be able to incorporate what you found into your writing. Data scientists can be an incredible resource in situations like this if you have access to one.
- **Coworkers:** Depending on where you work, chances are, there's a team or two that talks to your customers on a regular basis. Maybe it's the sales team. Maybe it's customer support. In any case, talking to those coworkers will not only make them feel appreciated, but will also teach you about your users and the language they use.

When working with customer data like this, take care to do it ethically. You and others don't need to know the names of the people involved or be privy to their private information. Learn about your users—don't spy on individuals.

This is a critical step, because you may find that the language your users use is quite a bit different from the language your team was planning to use. You can help your team see things from your user's perspective.

---

2 Sarah Richards, *Content Design* (London: Content Design London, 2017).

3 Lou Rosenfeld, *Search Analytics for Your Site: Conversations with Your Customers* (New York: Rosenfeld Media, 2011).

Writing with Plain Language

Most of the time, if your product is for the general public, you'll find that the best course is to use plain language.

That's harder than it looks: Unless your company has a profoundly user-focused culture (and let's face it, almost every company could be a bit more user-focused), it's easy to get caught in a bubble of business and software development jargon. From customer service systems that ask if your issue has been resolved (do real people ever ask each other to "resolve issues"?) to technical errors that are only meaningful to the people that wrote them, there is plenty of opportunity to make language more understandable and simple.

Maybe you've managed to stem the tide of buzzwords and filler phrases like "go ahead and . . . " and "at this time . . . " but there are still many terms and phrases that won't make any sense to the user.

The Nielsen Norman Group conducted a usability study showing that plain language benefits all readers:

- It's concise, helping users understand a concept quickly.
- It helps those who speak English (or whatever language you're writing) as a second language.
- It makes your language more searchable and improves your search engine optimization (SEO).

These points can help you convince marketers and decision-makers to simplify the words in product marketing and onboarding experiences.

Using Jargon: Your Results May Vary

There are so many different kinds of software in the world, and the threshold for plain language can vary wildly depending on the app. Are you designing something for a very specialized audience, like construction equipment purchasing managers? Or are you designing for the general public, like an event-ticketing app? If UX writing is a brand new discipline at your organization, and you don't have the capacity to go deep on domain expertise (don't worry, neither do most of us), you may have no insight into what words your users understand and use.

Plain Language in the U.S. Government

It seems hard to believe, but governments are doing some of the best work when it comes to writing in plain language. In 2010, the United States federal government passed the Plain Writing Act that provides a clear definition of plain language:

> Writing that is clear, concise, well-organized, and follows other best practices appropriate to the subject or field and intended audience.

It also, just as importantly, mandates that all government offices and services write plainly and according to their guidelines.

And while different offices and administrations have achieved varying levels of success with using plain language, there's one in particular that builds it into their DNA.

18F is an organization within the General Services Administration of the U.S. government that builds products, collaborates with other agencies, and serves with a mission to deliver great, useful digital products to the government. In addition to exceptional products, they create and publish standards and guidelines. The *18F Content Guide* is one such document that can help with many products and organizations well beyond the government.

One of the best pieces of advice their guide gives is powerful and should be the principle underlying everything you do: **"Do the hard work to make it simple."**

Here's what they say about that:

> Help the reader follow along. Break instructions or processes down into individual steps. Use short, simple sentences with words people use in everyday conversation.
>
> Refer to navigation labels, buttons, and menus as they appear in the app or website. Verify the spelling and capitalization as you write. Be specific.

Instead of:

> Open a new meeting invitation.

Use:

> In Google Calendar, select "Create."

Facebook, for example, takes great pains to distill their language into the simplest form possible. That's evident in the names they give the core components of their ecosystem, like "Pages," "Groups," and even complicated, advertising metaphors like "conversions" and "engagements." This makes perfect sense. They have a user base of more than two billion monthly active users, and their software is localized and translated into more than 100 languages.

But as a word of warning: Make sure that the jargon you do use is truly widely accepted and used in your users' industry. As we discuss in Chapter 2, research is a great way to find out how professionals talk. Sometimes, a short conversation with professionals about their work is invaluable to writers and designers alike who are trying to build products for them.

The best way to solve this would be to put the shortened copy, along with a prototype, in front of real users! You'll learn pretty quickly if that message confuses them, or if they have enough context clues to make an informed next step. (We'll talk more about user testing in later chapters.)

GOOD JARGON

Plain language is almost always the right way to go, but there are exceptions.

I once designed software that modeled machines and helped engineers size and select the components they needed to build it.

For example, they could be building a conveyor belt that moved tomatoes up an incline, through a washing mechanism, then placed them individually in cartons and labeled them. This software would allow the engineer to enter factors like the angle of the incline, the load on the conveyor belt, and the required speed, and then size the right motors and drives needed to build it.

In this case, it wouldn't make sense for me to use the interface to explain how load affected incline or what those terms meant. In fact, that would just have slowed them down as they used the tool throughout their workday. In the end, this interface didn't make sense to anyone but its users. However, it helped them size the parts they needed quickly and easily. ■

IN PRACTICE

Simplifying the Language of a Confirmation Dialog

Let's look at a warning dialog from an interface used by companies to assign software access to users. In this scenario, the admin (administrator) is turning off the feature that allows their users to share links to their work publicly—perhaps the company's privacy policy changed, or someone shared a link that wasn't meant to be shared and the company is cracking down. But when they turn off that public link-sharing feature, they may not realize that all previous publicly shared links are turned off retroactively. Figure 3.4 confirms the consequences to the administrator of taking this action.

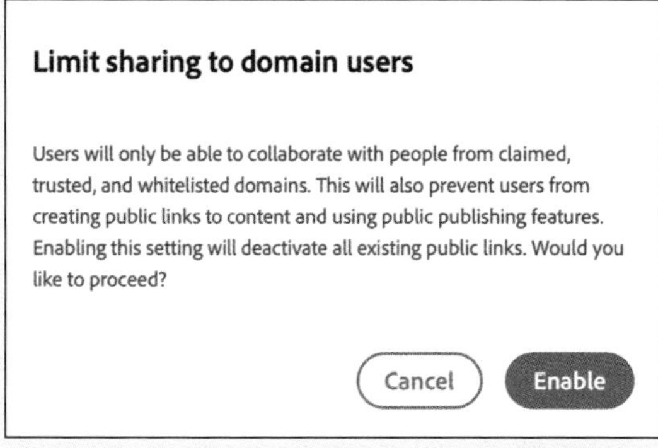

Limit sharing to domain users
Users will only be able to collaborate with people from claimed, trusted, and whitelisted domains. This will also prevent users from creating public links to content and using public publishing features. Enabling this setting will deactivate all existing public links. Would you like to proceed?
Cancel
Enable

FIGURE 3.4 An example dialog from an enterprise software company's admin console warning the admin user about a complicated consequence of changing role permissions.

This is a mouthful. Let's say that you're trying to edit this warning down to be more effective and to lighten the cognitive load on the user. It's complicated, but you can start by breaking down the copy into the bits of information included. This really gives you a deeper understanding of the message that's being conveyed, front and center. You're probably already doing this in your head, but let's do it on paper:

- The action you're taking will limit sharing to users in your organization.
- Users can only share and collaborate with people from certain approved organizations (represented by "domains").
- Users will not be able to create public links to their work.
- If you turn this setting on, all existing public links will be deactivated.
- Are you sure? Yes / No

This lets you sit back and really soak in the message. Look at all these messaging points you're trying to hit into one confirmation dialog.

continues

IN PRACTICE (continued)

Dig into the Flow

Since writing is designing, you really can't just edit this message in isolation. You need more context around where a user is coming from and what is prompting this message to appear. If you don't have access to the prototype or a design file for this message, ask the designer or product owner you're working with. It's important to see what the user saw before this message. Figure 3.5 is the list of sharing options the user saw before they got the confirmation dialog in Figure 3.4.

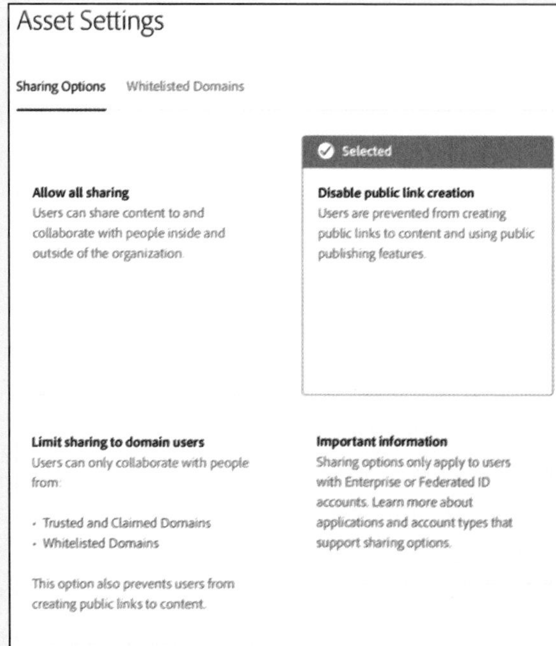

Asset Settings
Sharing Options
Whitelisted Domains
Selected
Allow all sharing
Users can share content to and collaborate with people inside and outside of the organization.
Disable public link creation
Users are prevented from creating public links to content and using public publishing features.
Limit sharing to domain users
Users can only collaborate with people from:
· Trusted and Claimed Domains
· Whitelisted Domains
This option also prevents users from creating public links to content.
Important information
Sharing options only apply to users with Enterprise or Federated ID accounts. Learn more about applications and account types that support sharing options.

FIGURE 3.5 A settings screen that allows the admin to set the sharing privileges of the users in their organization. Currently, the option that disallows public sharing is selected.

Prioritize the Message

Now that you understand a little bit more about this interaction, you can decide which information is more important for the user to see by re-ranking your messaging points. In this case, it's really important that the admin knows that any existing public links will be turned off if they proceed with this change, so let's move that from the bottom to the top of this message.

Put the Question First

Next, see where you can consolidate. Since the *confirmation* of this confirmation dialog is really why you're here, reflect this in the title since that's the most important.

> Limit sharing to domain users?

And because UI is a conversation, you need answers for the user to give that reflect the question being asked in the title: your button set could more accurately reflect the question being asked as a simple "Yes" and "No" instead of the jargony and mismatched "Enable" and "Cancel."

Trim the Fat

Next, look at that first line of text in Figure 3.4, explaining what this sharing restriction means. Users just saw this in a previous screen, as seen in Figure 3.5. And since this confirmation isn't about education around types of domains, you can trim that out.

De-jargonify

And finally, there's some jargon we can trim out to simplify the language.

- Enable = Let
- Deactivate = Turn off

Until finally, you're left with a concise, clear confirmation dialog, seen in Figure 3.6.

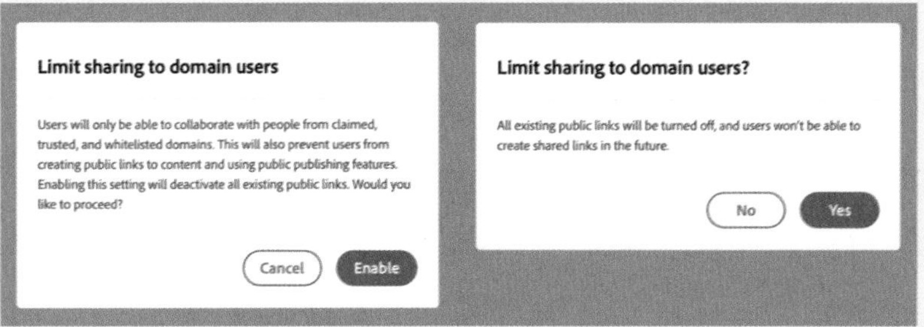

FIGURE 3.6
The original confirmation dialog, left, with the edited dialog, right.

It's shorter, by more than half, and much easier to read. This really lets the message stand out and be clear about the consequences of taking this action.

Finding What's Right for You

The way you achieve clarity can vary widely, depending on so many things—the functionality of your app, the type of user you're targeting, the context they'll have when using it, and so much more. Just as a serial comma doesn't make sense in short form journalism, but is completely necessary in legal writing, "clarity" is a moving target. As with *everything* in this book, it's super important to test your messages and interfaces with users.

CHAPTER 4

Errors and Stress Cases

When Things Go Wrong

Error messages are one of the most highly visible and top-of-mind examples of UI writing out there. They come when users are expecting something else to happen, and often cause frustration and confusion. They're usually the bearer of bad news—*Oops, we can't do that,* or *Sorry, something went wrong*, so by their very nature, there's no "good" error message. The best error message is no error message at all.

To understand errors, you have to understand how most people interact with digital products and services. They're not thinking about the website, mobile app, voice system, or other interfaces you created to help them do it. Their first priority is what they came here to accomplish.

Here are some typical things they're trying to do:

- **Look something up:** A parent with a sick child may need to look up common flu symptoms or find the nearest pharmacy.
- **Buy something:** A grandparent with poor eyesight may be shopping for baby clothes for the newest member of their family.
- **Get something done:** The person who pays their household bills may need to update the bank account number their utility company charges.
- **Create something:** A student may be using a combination of tools to create illustrations for an upcoming presentation.

As people try to complete the task at hand, inevitably, they'll run into negative moments, like error messages and warnings. These moments make it harder for them to get the thing done.

In almost every testing session we've led or observed, users feel bad when they encounter an error—frequently apologizing for using the software wrong. If this happens enough, this negative feeling boils over, and people give up completely.

But they're not using it *incorrectly*, they're just trying to use it. If people are frustrated because of errors, your response should be to design a better experience for them.

Before you start writing error messages, reframe error states as opportunities to help people get more done. Kathy Sierra does that in her book *Badass: Making Users Awesome*:[1]

---

1 Kathy Sierra, *Badass: Making Users Awesome* (Sebastopol: O'Reilly Media, 2015), p. 27.

> Where you find sustained success driven by recommendations, you find badass users. Smarter, more skillful, more powerful users. Users who know more and can do more in a way that's personally meaningful.

Rather than thinking about these things as errors, your team will be far better off if you focus on helping your users.

The Blame Game

Users aren't alone in bearing the error on themselves—the companies making the experiences often fall into the habit of blaming them too, even when the technology itself or the way it was designed is what's actually failing.

Companies do this because they feel they have a brand image to protect. Unfortunately, passive aggressive messages that make your users feel bad also hurt your brand image, while lowering user adoption and retention—two metrics that are important to the growth and development of a product.

If you find your team concerned about this, here are some questions you can ask your team and talk through with them:

- **What could cause this error?** There's usually a lot more that could go wrong than user input, from technical failure to unique user scenarios.
- **In what real-life context might the user be encountering this message?** These messages don't exist in a vacuum. Consider things like the user's physical environment or the time constraints they might be facing.
- **Can we test this?** Staying objective and letting your users do the talking can often be the most effective approach to provide guidance on how to frame this message.

It won't always be a problem caused by your business or your technology, but when that's the case, it's not such a bad idea to embrace authenticity.

Slack, a popular messaging service for workplaces, became a darling of the design community and of its users by doing just that. In Figure 4.1, what Slack described as their most popular error message of all time demonstrated what a company could do by shouldering the emotional burden of something going wrong.

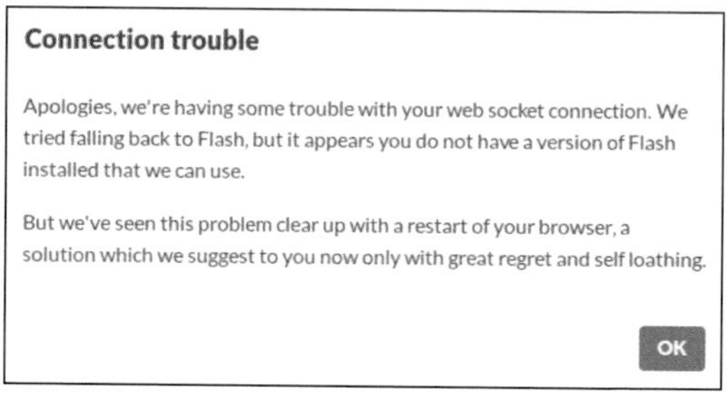

FIGURE 4.1
An old error message from Slack, taking the blame for when a web socket connection failed when loading the app.

Connection trouble

Apologies, we're having some trouble with your web socket connection. We tried falling back to Flash, but it appears you do not have a version of Flash installed that we can use.

But we've seen this problem clear up with a restart of your browser, a solution which we suggest to you now only with great regret and self loathing.

OK

It's wordy, but it's a great example because Slack took a frustrating situation and turned it around, helping users move forward toward their goal of connecting to Slack. This example goes to the extreme, but because Slack's brand carries a strong voice, and the context in which they present their error messages can accommodate so many words, it works.

If something goes wrong, do everything you can to make sure that your users aren't taking the blame. Focus on helping them achieve their goal.

Design by Discovery

Rather than diving right into the writing, start by asking questions about the errors:

- **Why does this scenario exist?** Is it a business policy? A legal requirement? An interaction quirk? If you don't understand why it's there, you'll have a hard time writing it.
- **What happens before this?** Context is crucial to understanding how a user may move off the "happy path" your software has designed for them. Learn as much as you can about what your users are going though. Look for existing research or do your own if you need to.
- **What's causing the error?** An error state could be triggered by something a user has done, a malfunctioning technical system, or a constraint of the software. Whatever the case, you won't be able to help your users understand what's going on if you don't understand it yourself.

Asking questions may not feel like part of the work, but it's just as important as the writing. Doing this type of discovery makes the rest of your work far more informed, effective, and efficient.

Lauren Lucchese learned this early in her career. She's a design leader and writer who has worked on financial products, artificial intelligence-driven products, and ecommerce—all while focusing on the words. While working on a financial product, one of her first assignments was to write error messages for the sign-in screen. She was given a spreadsheet with over 50 error codes and asked to "write something generic" that could apply to most, if not all, of the errors. Wanting to be helpful, she started to write, but soon realized she needed a different approach.

Lucchese started asking questions and found that generic error messages for this scenario could only lead to two results: instruction to call customer support, or a complete dead end. Obviously, neither was ideal.

On the other hand, messages that were specific to the customer's situation could provide enough guidance for them to solve the issue on their own, in some cases, without having to call. For example, if they were simply entering an incorrect password, explaining what was happening and sharing some of the password constraints could help them troubleshoot and move on more quickly.

For other errors, like suspected fraud, the customer would still have to call. However, the message could show consideration for their security, while providing a direct number for the fraud department. This way, they could speak to someone as soon as possible without having to go through the added frustration of a phone tree, especially at a stressful moment. A similar approach could be applied to the situation when an account holder was deceased. Whoever was trying to sign in to the deceased person's account could be sent directly to a representative equipped to handle the issue.

To deviate from the generic approach, though, she had to get agreement from her product owner. On digital products teams, product owners are often responsible for making sure the business gets what it wants, and often, the business wants your team to get things done quickly.

"When you go to a product owner and say 'Actually, this is way more complicated than you're making it out to be, and unless you think about all these things, you're not going to get the outcome you want,' that sometimes doesn't go over well," Lucchese said. "That's where you start thinking about metrics and measurements. You have to monetize the risk of not considering these things."

To sell the need to do more work up front for customers, Lucchese had to build a business case for why this approach was viable. She knew a key goal was to lower the number of calls to the call center and reduce call time, both of which cost the company lots of money.

She worked with the analytics team to get data on how many people called because they were locked out of their accounts and couldn't figure out why. She also collected data on how long these customers stayed on the phone with a representative (some calls lasted up to 30 minutes!) and what the associated costs were. She used this data to build the case for more specific error handling.

Soon the whole team realized the business value of delivering the right message, and they were able to see the results.

"We used some really targeted error messages to help clarify what we knew would be sticking points, regardless of how we set up the page," she said. "It made a huge difference."

These updates, along with other improvements to the onboarding experience, contributed to significantly higher sign-in and adoption rates. Lucchese and her team ended up creating a system that was far more valuable to the business and far more appropriate for the people who were using it.

Writing Error Messages

Once you understand what your users are going through, you'll start to see what error messages are needed and when they'll appear. But how should you write them?

Here are three principles for using words to design error states—one of them doesn't even involve writing:

- **Avoid:** Find ways to help your user without showing them an error.
- **Explain:** Tell your users what's going on and what went wrong.
- **Resolve:** Provide a solution to the problem that the user is facing.

To see how this plays out in the real world, let's look at an increasingly popular technology as an example: online check deposits through a banking app.

Paper checks are alive and well, but now you don't even have to leave your home to add funds to your account. This feature is so popular that according to Bank of America's CEO, more checks were deposited through their mobile app than in local branches from April to June of 2018.

If the following conditions aren't met, the check-cashing feature could generate an error:

- Users must enter an amount for their check.
- The amount entered by the user must match the amount shown on the check.
- The image of the check must be clear enough that the software can read the account numbers, routing numbers, and amount.
- The check must be signed.
- The user must write the words "For online deposit only" on the back (a legal requirement as of 2018).

These are just a few possibilities! There could be many, many more situations to account for. Let's see how these concepts play out in the real world.

Avoid

Avoiding errors is really about preventing frustration, while helping people accomplish their tasks. Ideally, it leads people to deposit their checks without any problems, allowing them to move on with their lives. The fewer issues they face, the more convenient an experience it is for the customer and the lower the cost for the bank.

The best experiences reduce the need for errors by using visual cues and interaction patterns to guide their users. Figure 4.2 shows the Chase Bank mobile app's check deposit feature. The user must enter the amount before taking a photo of the check, so the designers made that input stand out in large text and grayed out all the other options, guiding the user to enter the amount, before making the photo-taking buttons available. If users could select the "Next" button without entering an amount, they would encounter an error state.

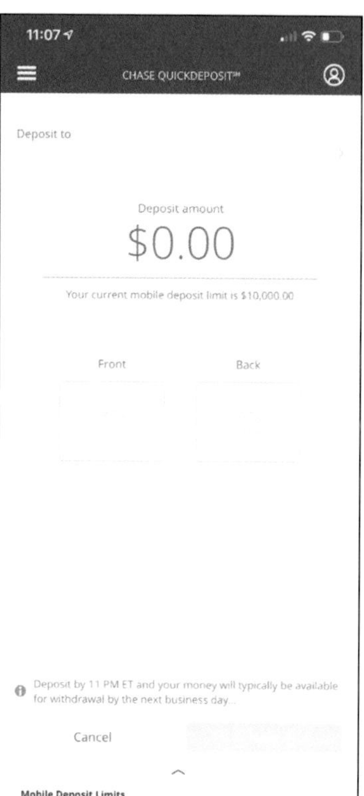

FIGURE 4.2
By using visual cues and disabled states to guide users, this part of the Chase mobile app's check deposit feature doesn't need an error message.

Another approach to avoid an error is to design the software so that it assists users with formatting and data entry. Some systems require users to enter their birthdates, and that data is required in the mm/dd/yyyy format. To prevent errors, the forms can be created to add the slashes automatically. You could also use a date-picker, which allows a user to choose the date from a calendar or a "wheel" of years, days, and months, but there are speed trade-offs to consider. Someone may have to click or tap through many years to find the one they were born in.

Conversational interfaces like voice assistants and chatbots take similar steps to reduce errors. If you're writing dialog for one of these interfaces, you'll likely want to hint at the correct way to enter information. If you've ever called an airline's automated phone system and heard it tell you, "You can say things like 'check the status of a flight' or 'change my reservation,'" the system's designers are working to reduce errors before they happen by helping users understand the system's capabilities and what types of commands it responds to.

There are many, many ways to avoid or reduce errors before they happen. It takes a design mindset, a strong understanding of the technology involved, and a deep familiarity with the business constraints to do it effectively.

Explain

Back to depositing checks: There are cases where you won't be able to avoid an error message. Let's say one of your users gets a large check (lucky them!), and they enter the amount. In this case, the bank has imposed a "mobile deposit limit" of $10,000.

Figure 4.3 shows how the Chase app handles this scenario. In these cases, the idea is to tell the user clearly and quickly what went wrong. The Chase app does a decent job here, but it would fall short if the user wanted to know why the deposit limit was $10,000. They might also want to know where the check could be deposited. It's implied that this is a "mobile" deposit limit, but does it apply to ATMs, too?

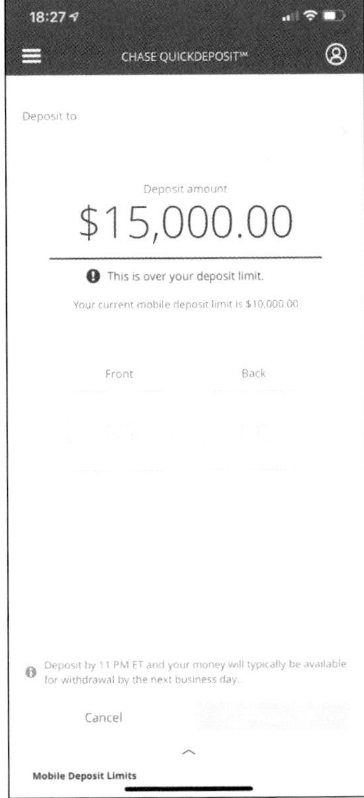

FIGURE 4.3
The error message for when the user exceeds the deposit limit is short and clear, but it doesn't help the user understand what to do next to deposit their check.

But that information isn't here. Why? Maybe there was a limited character count for the error field. Or maybe the business didn't feel it was necessary. Or it could be that the design team just didn't think of including it.

Again, it's critical that you foresee and solve problems before they happen, and by doing that, you're treating writing like what it is—design. If you think of your contribution as writing a message for a specific field with a limited character count, you're limiting the ways you can help your users. No matter what your title is, you can work with your team to explore options like the ones shown below for approaching the problem differently:

- Adding a tooltip (a small message that appears when you click or tap an icon)
- Increasing the character count of the field, or if you're building from scratch, collaborate with the team to figure out what the character count should be
- Adding a link to an explanation of this limit (websites are great for that)

The best way to figure out the right level of detail for an error message is to let your users tell you. You can test the words with users just like any other aspect of the interface.

Showing solutions is what ultimately builds trust with your users. When you fail to explain what's going wrong, it can mean the difference between a user embracing or rejecting your product.

Resolve

If, after following both of the previous principles, you still have to write an error message, then resolving the issue that led to the error is easily the most critical component—for both your users and the business. It's one thing to explain what's wrong, but you certainly don't want to leave the user asking, "Now what?" and abandoning the product entirely.

Resolving is about offering next steps. Often, this involves backing up and understanding the user's intent.

Figure 4.4 shows what happens when the Chase app's "auto capture" feature isn't able to capture a clear image of the user's check. "Auto capture" takes a photo of the check by detecting whether it's in the frame, and then it triggers the camera.

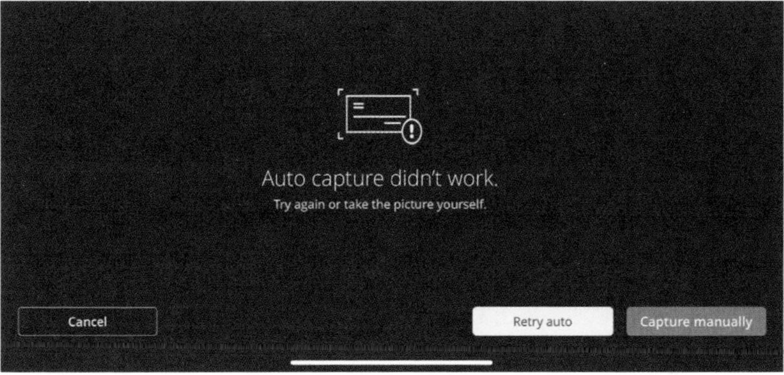

FIGURE 4.4

This error state gives the user two options to move closer to accomplishing their goal.

This is more than an error message. "Capture manually" and "Retry auto" give the user two ways to resolve the issue and move closer to their goal of depositing a check. "Take the picture yourself" might come off as bossy, but testing would tell you whether users perceive it that way.

Resolution is about meeting users where they are and providing a path forward. It's about helping users get things done.

DEALING WITH COMPLEXITY

Sometimes, helping users resolve the problem they're facing is a challenge, because it means rethinking how the system was designed and built.

I ran into problems with resolution while designing a chatbot. Its purpose was to educate users about many complex topics. However, we were just at the beginning of building and testing the product, so it couldn't answer very many questions.

When users asked a question, the bot would respond with something from its built-in error pool. Things like:

- I don't know.
- I'm not sure about that one.
- Sorry, I didn't catch that.

It was to the point where users were getting an error message the majority of the time, and these messages made them feel like it was their fault—as if they were using the product wrong.

Initially, I was asked to rewrite the error message, to solve this problem, but just saying "I don't know" in a different way wouldn't help our users resolve the issue. They would still abandon the interaction in frustration.

We ended up building in logic that could separate the first time the system didn't know the answer from the second time, and help users move closer to resolution each time.

The first time the system didn't know the answer, it would say:

> Sorry, I don't know. My knowledge is limited right now, but rephrasing your question may help.

The second time a user encountered the message, it would be different:

> I don't know the answer.
>
> I can't help, but your questions are helping me learn. Would you like me to transfer you to a support analyst?

The reason this was so hard to do was that we still had to use that default "error" pool. It took the engineering team modifying the underlying platform to make these conditional messages work, but it was worth it.

This move toward resolution had a clear impact on getting users to use the product more, because users went from being discouraged to believing that they were helping the product while getting the support they needed. Ultimately, we were able to increase adoption significantly just by focusing on error resolution. ■

Test Your Message

Even if you do a great job discovering user needs, preventing unnecessary errors, and writing the messages so they explain the problems and help users resolve them, you still may get something wrong.

How can you know for sure? By testing those words with your users.

Natalie Yee is a UX designer who has been exposed to a wide variety of work, but her experience with medical products is what has shown her the value of testing the words in an interface.

For one project in particular, she described how the team uncovered some of the reactions that users were having to a situation where emotions and stress levels often ran high.

"We printed out the screens and just put them in front of people so there was no way to interact with the app," Yee said. "A lot of people just showed emotion right away and would say things like 'That would make me mad.' or 'That's fine. I would wait.'"

One feature has a limit on the number of times it can be used each day; then it would show an error message. The message would say they had reached the limit, and that they could use the feature again the next day.

Users thought something was wrong with the system when they saw this message. In fact, it gave them the impression that their information was being lost. In reality, the information was still collected, regardless of whether they encountered the error.

"We updated the error message to explain that to the user," said Yee. "When we tested it again, the error message was reassuring instead of concerning."

The testing showed the entire team that they could handle some of those situations differently and improve the experience as a result.

"We actually changed a lot of text from those tests because we realized the messages were alarming, or that people would have no idea what to do next," she said.

You can test any product and any message. As Yee explained, your users will be glad you did. Yee's experience testing error messages gave her a clear perspective on why words were so important: "We know that words can really hurt people or help them in their personal lives," she said. "We can say really reassuring words to people, and it has this huge impact, and we can say hurtful words, and it can have a years-long impact, but we don't really treat the words we write in interfaces that way."

By spending time on the words, Yee keeps the focus on her users and what they're going through.

"It's important that the copy you write aligns with the rest of the app, but it's more important that you think about the impact your words can have on people's lives outside the app."

WHEN THE WORDS AREN'T THE PROBLEM

"We need you to write an error message for when people are over the age of 100."

I had just arrived at work when the engineering lead I'd been working with delivered this to-do.

"Huh?" I asked, if you can call a grunt "asking." I was barely awake at this point.

"The sign-up form asks for birth date, but the security guidelines say that we can't accept any birth dates more than 100 years ago. Can you write an error message? We're working on integrating the sign-up form with the database right now. We put in 'Please enter a valid birth date' as a placeholder."

I had so many questions.

Why does this security guideline exist? Do other forms at this organization throw a similar error? How could I possibly write a message that explained this policy?

"So if someone is 100 or older, we have to throw an error on the sign-up form?" I was finally waking up.

"Yes," she responded.

At this point, other people in the room, like our product owner and visual designer were starting to notice this conversation.

"Will we really get that many people who are over 100 signing up for this?" the product owner asked. It often falls on them to find a way forward in these situations. They were asking because accommodating these users was going to take time. Time they felt we didn't have. Besides, wasn't this kind of an edge case?

Sometimes, you'll encounter an error message that will be impossible to write. That's because sometimes the policies driving the error aren't human-centered, they're business-centered.

In their book, *Design for Real Life*, Eric Meyer and Sara Wachter-Boettcher make the point that the term "edge cases"—frequently used by the teams that build software to describe a small subset of users—is designed to make it seem like those people don't matter. "Don't worry," teams say, "that's just an edge case."

Instead, they prefer the term "stress case," because it draws attention to the frustration and emotion these users might feel when interacting with the product.

While the members of your team may not meet people who are 100 years of age or more each day, that doesn't mean they don't exist. In fact, the population is growing. A 2010 census showed that there are 53,364 people in the United States alone who are 100 or older, up almost 65.8% from the 1980 census.

If a sign-up form throws an error whenever one of these people attempts to purchase the product, we're telling 53,364 people that we don't care for their business—and that's the best-case scenario.

Anyone working on a product can write error messages, but if they're your responsibility, you're the person best equipped to start conversations around whether those words meet the needs of your users and your business. If you're designing, and not just writing, you'll end up solving problems like this and making a better experience for *everyone*, not just the majority of your users. ■

Finding What's Right for You

Chances are, when most people think of "design," they're not thinking of writing an error message or figuring out what a product's stress cases are. However, they're more than messages. They're critical moments that can make or break the user experience.

Since writing is designing, each error is an opportunity to help your users get things done and deliver a product that works for everyone.

CHAPTER 5

Inclusivity and Accessibility

Writing That Works for Everyone

Inclusive experiences benefit everyone. About 20 years ago, Matt May learned this while he was working for an online grocery-ordering website based in Seattle, and worked on a service intended to help busy parents save time on shopping.

"We started getting calls from [disabled] people saying, 'While you're focusing on soccer moms saving time buying groceries, it's a four-hour process for me because I have to call the grocery store, tell them what I need, and then they tell me what they've got. Then I have to book an access van to take me, all because I'm blind, or I have a disability.' I realized that the 15 minutes a day I was saving for [abled] people's convenience was saving others the better part of a day."

That really clicked for May and influenced his career. He went on to work as a web accessibility specialist for the World Wide Web Consortium (W3C), and later coauthored a book, *Universal Design for Web Applications*, all about building accessible web products. He's now Head of Inclusive Design at Adobe.

There's a lot of overlap between "accessible" design and "inclusive" design, and we'll be covering some of both in this chapter, but they are definitely different.

This is a question May gets a lot. "Accessibility is the destination, the goal, the objective," he said. He explained that access is about filling in the gaps in usability for a disabled population.

The word "access" has other meanings. Some people might say, "Oh, the website is accessible 24/7, so *of course* it's accessible to everyone." We already face a terminological block in people's heads when we assume users know what it means.

"So one of the reasons I think that 'inclusive design' is a useful differentiator, is that it's broader, and encompasses more than disability," May said. "Sometimes that's disability. Sometimes it's race. Sometimes it's age. Sometimes it's gender. And sometimes, they think about their own lived experiences. All that is valid, and needs to be part of the discussion."

Basically, May says, if "accessibility" is the port your ship needs to arrive at, "inclusive design" is pointing your ship toward that port when you set out to sea.

Making the Case

Inclusive language helps everyone feel like your product is made for them, but you may work with some people who object to the idea that you're spending time and effort making your product inclusive. They may say, "This experience works for 95% of people. Isn't that enough?"

Well . . . no. At the time of this writing, there were 7.5 billion people in the world. If you exclude even one-tenth of one percent, that means there are 755 million people who are less able to (or can't) use your product, or pay for your service, or experience your interface.

In reality, that number is much higher. In 2018, the World Health Organization reported that there were an estimated 217 million people with a severe vision impairment—36 million of those people were blind. And 466 million people had a disabling hearing loss or deafness.[1]

In the United States alone, the Reeve Foundation estimates that around 5.4 million people have paralysis of some kind.[2]

When you think about your user base's gender and sexuality, there are so many people you often and easily exclude. The Williams Institute, part of the UCLA School of Law, estimates that 10.3 million adults in the United States alone identify as lesbian, gay, or bisexual—and 1.3 million identify as transgender.[3]

All this is to say, those numbers add up. And if the language you use tells a potential user that *you didn't think about their experience when building this thing,* they're not going to use it. That's a big chunk of revenue you're leaving on the table.

Being thoughtful and inclusive in the language you use isn't just about being a social justice warrior (although what's wrong with social justice?) or politicizing your work. It's good for business.

1 "Deafness and Hearing Loss," World Health Organization, published March 20, 2019, https://www.who.int/news-room/fact-sheets/detail/deafness-and-hearing-loss

2 "Stats About Paralysis," Christopher and Dana Reeve Foundation, June, 2019, https://www.christopherreeve.org/living-with-paralysis/stats-about-paralysis

3 "Adult LGBT Population in the United States," UCLA School of Law Williams Institute, March 2019, https://williamsinstitute.law.ucla.edu

Inclusivity

In her 2018 book, *Mismatch: How Inclusion Shapes Design*, author Kat Holmes talked about how designers and writers often design and write for themselves when creating products. And even when they keep accessibility in mind, there's a spectrum of people using their product throughout who might not have a great experience.

The *Microsoft Inclusive Design Manual*, which Holmes helped develop, sums it up well: "Ideally, accessibility and inclusive design work together to make experiences that are not only compliant with standards, but truly usable and open to all."[4]

"One of the greatest shortcomings of human-centered design is its lack of guidance on how to include diversity as part of the design process. Which human, exactly, belongs in the center? Most designers end up using their own abilities and experience as a baseline for their designs," Holmes wrote.

The Microsoft manual gives some great examples of this. For example, when you create a touchless interface, like a smart speaker, designed for someone with limited or no mobility in their arms, you're also helping someone with situational mobility disabilities—a new parent who needs to navigate products and websites with a baby in their arms, for example. An interface that's designed with deaf people in mind also helps someone who may have limited hearing because of an ear infection, or even someone who is in a crowded or loud place, like a bartender.

The manual says that all of us experience disabilities in one way or another over our lifetime. By focusing on the universal ways that humans experience the world, designers and writers can magnify their impact of inclusion exponentially.

AN EXTRA MILE FOR DEAF DRIVERS

My friend Lauren recently got a notification from the ride-sharing app, Lyft, after being matched with a driver. It told her that her driver was deaf or hard of hearing (see Figure 5.1), and that if she needed to get in touch with them when they were picking her up, she should text instead of call.

4 "Inclusive Design," Microsoft Design, June 2019, https://www.microsoft.com/design/inclusive/

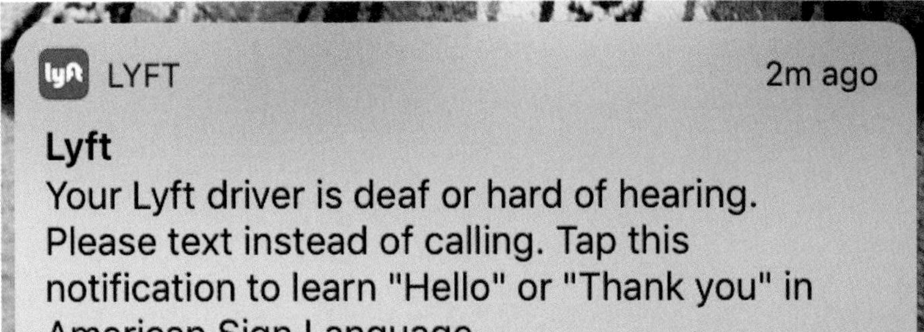

FIGURE 5.1
A push notification received by Lyft passenger Lauren Caggiano, telling her that her driver was deaf or hard of hearing, and that she should text them instead of call. Lyft even included a way for Lauren to talk to the driver in ASL if she wanted to.

I'm just speculating here, but the percentage of Lyft drivers with a hearing difference must be pretty low. This is a great example of Lyft practicing inclusion to make the experience for those drivers better. Rather than being flustered about answering a phone call they couldn't hear, they could pull over, text back, and communicate in a way more comfortable for them.

I've gotten similar messages to this before myself, and often drivers who don't speak English or aren't comfortable conversing during a ride will use this feature. It's not the way Lyft intended, but it's a great hack, because it signals to the passenger that they don't want to (or can't) communicate verbally. ■

Unpacking Fitbit's Pronouns

One person who sees how important it is for anyone building digital products to develop a more robust appreciation for ethics and equity is Ada Powers, a community organizer and UX practitioner. She thinks design is a crucial part of how inclusion can be instilled into the practices of big companies.

Powers is trans, and she sees exclusive language all over the place, woven throughout the products she uses. One example is in her Fitbit: when she was filling out her profile, the interface asked her to choose her sex: male or female. (See a screenshot of the interface in Figure 5.2.) It seems straightforward, right?

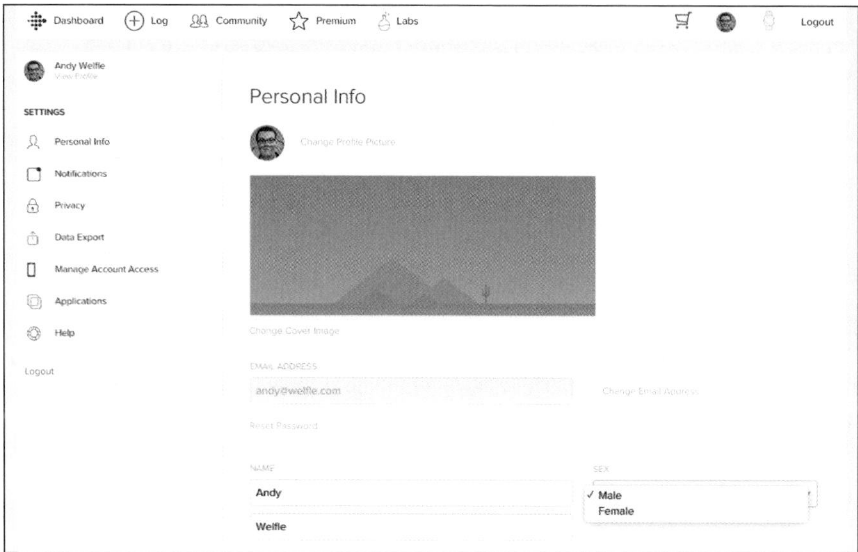

FIGURE 5.2
Screenshot of Fitbit's profile form.

For Powers and other trans folks, this question isn't straightforward at all.

"Now, I want to get the most out of my Fitbit," Powers said, "so I want to understand where they're coming from. Normally, I'd assume they're awkwardly asking about my gender and just answer 'female,' but they also asked me for my height and weight, so maybe they're looking for biological rather than demographic information. Should I answer 'male' because of my supposed chromosomes, or 'female' because the dominant sex hormone in my body is estrogen?"

"But it's even more complicated than that. If I answer 'female,' do they assume I'm interested in period tracking? What if I were a cis woman who doesn't have a uterus, or has one that doesn't function normatively? Do they use it to assume hormone levels for their calorie burning models? What about people with PCOS (polycystic ovarian syndrome), which can cause relatively higher levels of testosterone?"

Powers says that routine data collection like this often impacts her experience, because she's not privy to what they're doing with that information. How could they have made this workflow easier for everyone using their product? Powers says that transparency is key.

"If they were thinking about inclusivity, they'd understand that some questions don't have easy answers. By explaining what they want to know and why, it not only helps people on the margins like me, but anyone who may not be easily categorized—and gets them even better-quality information to act on. It's a win-win for the user and for the company."

Companies need to let their users know what they're doing when they collect this kind of information. Is it just for demographic reporting purposes? Is it for deciding what ads to show people? Or is it integral to the core function of the product?

It's hard, but it's important for a writer to question the collection of data in their products.

As that writer, here are a few things you can ask yourself and your product team:

- **What are we going to do with this information?** What sort of assumptions might we make based on this information?
- **How might someone misuse the information we're collecting?** If we're gathering demographic data that we're selling or otherwise making available to other people in the organization, how could someone use this information in an unethical way, even if it was against the will of your company?
- **How might asking for this information make someone's experience harder?** Who might experience stress or relive trauma because the information we're asking for excludes them from answering?

These questions might not make you popular with the business development folks, but it's worth building awareness around this issue and making the case for ethical data use. But if you are able to affect change and add transparency and context, you can help make an underrepresented user's experience easier.

You may think to yourself, "Well, this isn't such a big deal. It's just a choice on a form." And you'd be right. In isolation, it's a relatively small decision. But marginalized people face this tiny exclusion many—sometimes dozens—of times a day.

Think about Powers's experience with Fitbit. What if Fitbit's sign-up form had given context about why they wanted to know her sex? If it was for biometric reporting purposes only, or if it was for demographic reporting purposes, it would have been easier to make a choice. (Though, of course, the best way Fitbit could have handled it was to make it an optional choice, or to have given a free text field.)

Take a look at One Medical Group's patient intake form (see Figure 5.3), which gives more context around why they're asking for this information.

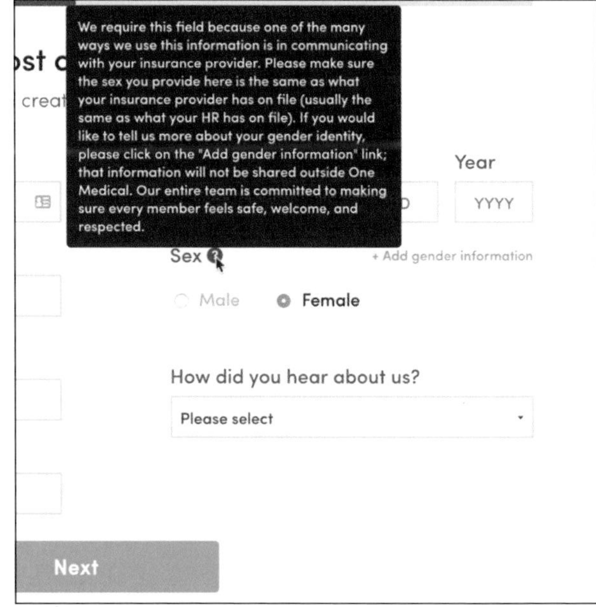

FIGURE 5.3
One Medical Group's explanation for why they ask for biological sex is transparent, although wordy.[5]

People and Identity

Context matters. When you're hanging out with a group of friends, you know how to talk to them, and you know how to talk about them. But when you're writing for users, you don't know who, specifically, those users will be. You don't know their lived experiences. You don't know their identity.

5 Sabrina Fonseca, "Designing Forms for Gender Diversity and Inclusion," *UX Collective*, April 24, 2017, https://uxdesign.cc/designing-forms-for-gender-diversity-and-inclusion-d8194cf1f51

Next, you'll find guidance on how to write more inclusively. However, here are two things you should keep in mind as you read it:

- **This is not a complete list.** There are a multitude of identifications in the world not covered below. The great irony of this chapter about inclusivity is that we're excluding plenty of people because of space constraints. This list is growing all the time as well. Our goal is to help you prioritize this in your own work so that you can continue to learn.
- **Thinking around this language changes over time.** Just as "retarded" was a medical term once used to describe the developmentally disabled that became used as an insult, it's now shockingly out-of-date. (You may bristle even as you read it in this context—we sure did.) As our society changes in how we treat underrepresented people, so should you in how you refer to them. Keep learning. Always.

Avoid Language That Assigns Value to Traits

While people-first and identity-first language doesn't have a defined rule set,[6] there is one pitfall you should definitely avoid: assigning value, whether it's positive or negative, to those with disabilities. For examples, look at Table 5.1.

TABLE 5.1 RESPECTFUL WAYS TO REFER TO IDENTITY

Avoid saying	She's confined to a wheelchair. She's wheelchair-bound.
Instead, say	She uses a wheelchair. (Remember, a wheelchair represents freedom of movement and access to its user.)
Avoid saying	He suffers from autism.
Instead, say	He has autism.
Avoid saying	They bravely overcame adversity by learning to talk again.
Instead, say	They regained their ability to talk after experiencing a stroke two years ago.

6 Don't just take it from us—the American Psychological Association at http://supp.apa.org/style/pubman-ch03-00.pdf has guidelines for writing about people clearly, concisely, and with as little bias as possible. The Linguistic Society of America also has great inclusive writing guidelines at https://www.linguisticsociety.org/resource/guidelines-inclusive-language

A DISABLED WORLD CHAMPION'S IDENTITY

My sister Nina, who is disabled and uses a wheelchair, is an impressive athlete and wheelchair basketball player. She trains hours each day, and it pays off. She recently competed with Team USA in the 2019 Women's World Basketball Tournament (see Figure 5.4) and her team won! Whenever someone in the media writes about her, they proclaim how brave she is, how she overcame all odds to become an athlete, and how inspiring she is for doing so.

Many disabled athletes find this patronizing and infantilizing. And so does Nina: "I hope that I'm inspiring because I worked really hard to be a good player," she said. "Not because I use a wheelchair."

FIGURE 5.4
Wheelchair Basketball player Nina Welfle with the 2019 World Championship Trophy.

And her dedication and work is inspiring. She's trained so hard for so long that when she was 12 years old (and I was 24), she could beat me at arm-wrestling. ■

Avoid Prescriptive Language When Talking About People

It's easy to make assumptions based on someone's experience or career, especially when you're making professional software. Your goal is to write words that are clear and relatable to *all* users, not just the ones who are paying or completely understand your product (see Table 5.2).

TABLE 5.2 DON'T EXCLUDE USERS WITH PRESCRIPTIVE LANGUAGE

Avoid saying	Adobe Photoshop is for graphic designers, photographers, illustrators, and 3D artists.
Instead, say	Adobe Photoshop helps you create graphics, photographs, illustrations, and 3D art.

The Singular "They"

It's okay to use the singular "they." And, in fact, it's preferred. Gender is not binary; an increasing number of people don't identify as a man or as a woman, and when you write, you shouldn't assume they are either. There's no reason to use the clunky "he/she" or even "(s)he" when you are writing using gendered pronouns. Instead, use *they, them, their, theirs, themselves,* and even *themself,* which are all acceptable.

And when the inevitable grammar prescriptivist at your organization chides you for using a plural pronoun to refer to a singular person, you can gently remind them that *they* has been used to refer to a singular person since the 14th century, not long after it was used to refer to plural people.[7] See Table 5.3.

TABLE 5.3 KEEP IT SIMPLE WITH "THEY"

Avoid saying	This form will be sent to a customer service representative. (S)He will be in touch within one business day.
Instead, say	This form will be sent to a customer service representative. They'll be in touch within one business day.

If you're using pronouns of your users in your interface, here are a few things to keep in mind:

- Ask for a user's pronouns separately from asking them their gender (though most of the time, you don't need that information at all). Don't assume that someone who chooses "male" uses "he/him" pronouns.

7 "They, pron., adj., adv., and n.," *The Oxford English Dictionary,* September 2013, https://www.oed.com/view/Entry/200700

- Don't ask for "preferred" pronouns. They are a statement of fact, not a request.
- Make that information editable. Assume that a user should be able to change their pronouns at any time.

If you work for a healthcare or health insurance company, you may be required by law to ask for a user's sex. Be sure to specify: Is it their sex assigned at birth? Is it their legal sex?

Read More About the Language of Inclusivity

The Conscious Style Guide[8] is a collection of articles, guides, and opinions on a wide variety of language that writers can take into account when writing about people to avoid exclusion. With articles spanning topics such as religion, sexual identity, gender, health, age, and ability, it's a great repository of the latest thoughts on inclusion and respectful discourse.

BEING INTENTIONAL ABOUT INCLUSIVITY

Let's get personal for a moment: the authors of this book are white, heterosexual, cisgender, middle-class men who live in urban areas of the United States—we have a lot of privilege and have never:

- Been misgendered
- Been harassed by the police because of the color of our skin
- Been passed over for a career opportunity because of our gender

...and much more. We've both written and designed with assumptions and biases, because there's an audience we know best—ourselves.

Designing for ourselves, and those with similar traits, is easy. Finding the words that best connect with ourselves, given our lived experiences, likes, and dislikes, is second nature.

But questioning your own assumptions and learning about the needs of others is critical as technology takes over so much of our lives. If writing and designing with an inclusive mindset is new to you, you're going to make mistakes. The important thing is that you respond appropriately and learn from them.

We've misgendered people, sometimes repeatedly. We've used the word "crazy" or "lame" in ways that makes light of those with mental illnesses or physical disabilities.

---

8 Karen Yin, "Conscious Style Guide," June 2019, https://consciousstyleguide.com

If someone calls you out on it (either personally, or through the lens of the words in your product), it's not a time to get defensive. Take the feedback as a learning opportunity and be transparent. Say, "I'm sorry" or "Thank you, I made a mistake and will work on doing better next time."

As you learn, take the opportunity to educate others you work with. It shouldn't be completely up to the people in underrepresented groups to bring these things up with the teams they work with, and making these issues part of the conversation creates a more inclusive tech industry. ■

Accessibility

Making your product work better for those with disabilities isn't just a good idea: for huge regulated industries like healthcare and public-sector websites, it's the law. In the United States, Section 508 of the Rehabilitation Act of 1973 requires all government websites and services to be accessible to people with disabilities. (That section was amended in 1998, in case you were wondering how Congress passed website regulation 16 years before the World Wide Web was created.) In the UK, the Equality Act of 2010 calls for a similar standard among government websites.

The standard by which writers and designers measure their content's level of accessibility is the Web Content Accessibility Guidelines, version 2.1 (WCAG2). We won't rehash those guidelines here (you should spend some time and look them up), but here's a summary.[9]

Perceivable

The content or operation can't be invisible to all of the user's senses.

- Provide text alternatives for non-text content.
- Provide captions and other alternatives for multimedia.
- Create content that can be presented in different ways, including by assistive technologies, without losing meaning.
- Make it easier for users to see and hear content.

9 "WCAG 2.1 at a Glance," W3C Web Accessibility Initiative, last modified June 5, 2018, https://www.w3.org/WAI/standards-guidelines/wcag/glance/

Operable

The interface cannot require interaction that a user cannot perform.

- Make all functionality available from a keyboard.
- Give users enough time to read and use content.
- Do not use content that causes seizures or physical reactions.
- Help users navigate and find content.
- Make it easier to use inputs other than keyboards.

NOTE BEWARE OF FLASHING VIDEOS

Epileptic people have to be careful with what they see on their devices, since flashing lights and bright, saturated colors can trigger a seizure. WCAG2 recommends avoiding videos that flash more than three times per second and to reduce the contrast of flashing colors.

Understandable

The content or operation cannot be beyond the user's understanding.

- Make text readable and understandable.
- Make content appear and operate in predictable ways.
- Help users avoid and correct mistakes.

Robust

As technologies and user agents evolve, the content should remain accessible.

- Maximize compatibility with current and future user tools.

Standards for Writing Accessibly

Writing to meet WCAG2 standards can be a challenge, but it's worthwhile. Albert Einstein, the archetypical genius and physicist, once said, "Any fool can make things bigger, more complex, and more violent. It takes a touch of genius—and a lot of courage—to move in the opposite direction."

Hopefully, this entire book will help you better write for accessibility. So far, you've learned:

- Why clarity is important
- How to structure messages for error states and stress cases
- How to test the effectiveness of the words you write

All that should help your writing be better for screen readers, give additional context to users who may need it, and be easier to parse.

But there are a few specific points that you may not otherwise think about, even after reading these pages.

Writing for Screen Readers

People with little or no sight interact with apps and websites in a much different way than sighted people do. Screen readers parse the elements on the screen (to the best of their abilities) and read it back to the user. And along the way, there are many ways this could go wrong. As the interface writer, your role is perhaps most important in giving screen reader users the best context.

Here are a few things to keep in mind about screen readers:

- The average reading time for sighted readers is two to five words per second. Screen-reader users can comprehend text being read at an average of 35 syllables per second, which is significantly faster. Don't be afraid to sacrifice brevity for clarity, especially when extra context is needed or useful.
- People want to be able to skim long blocks of text, regardless of sight or audio, so it's extremely important to structure your long-form writing with headers, short paragraphs, and other content design best practices.

Write Chronologically, Not Spatially

Writing chronologically is about describing the order of things, rather than where they appear spatially in the interface. There are so many good reasons to do this (devices and browsers will render interfaces differently), but screen readers show you the most valuable reason. You'll often be faced with writing tooltips or onboarding elements that say something like, "Click the OK button below to continue." Or "See the instructions above to save your document."

Screen readers will do their job and read those instructions aloud to someone who can't see the spatial relationships between words and objects. While many times, they can cope with that, they shouldn't have to. Consider screen reader users in your language. Embrace the universal experience shared by humans and rely on their intrinsic understanding of the *top is first, bottom is last* paradigm. Write chronologically, as in Figure 5.5.

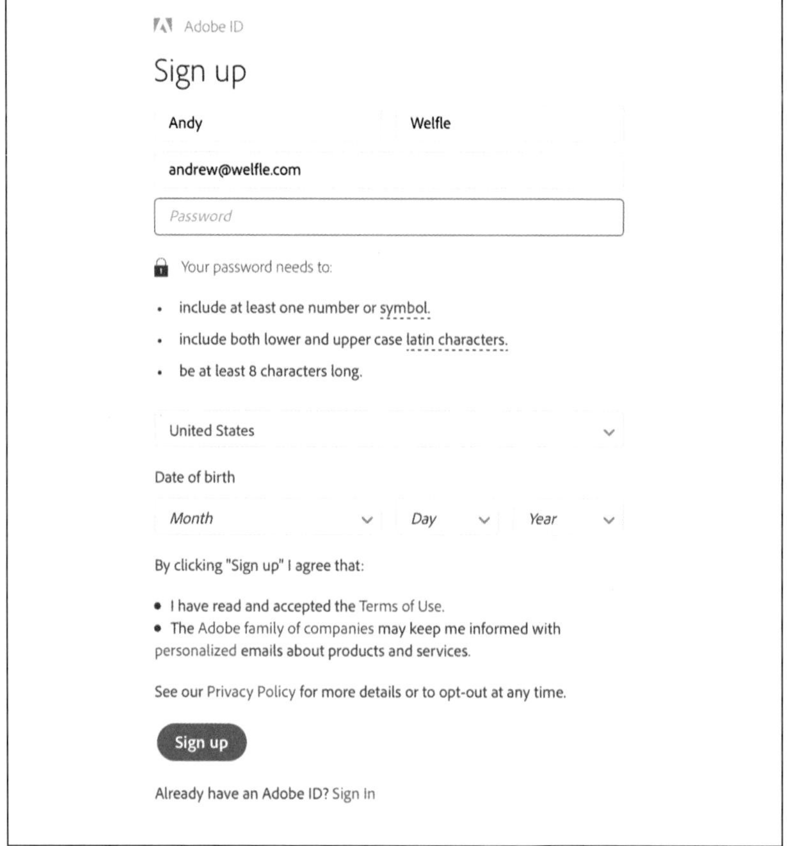

FIGURE 5.5
Password hint microcopy below the password field won't help someone using a screen reader who hasn't made it there yet.

Rather than saying:

- Click the OK button below to continue.
- (A button that scrolls you to the top of a page): Go to top.

Instead, say:

- Next, select OK to continue.
- Go to beginning.

Write Left to Right, Top to Bottom

While you don't want to convey spatial meaning in your writing, you still want to keep that spatial order in mind.

Have you ever purchased a service or a product, only to find out later that there were conditions you didn't know about before you paid for it? Maybe you didn't realize batteries weren't included in that gadget, or that signing up for that social network, you were implicitly agreeing to provide data to third-party advertisers.

People who use screen readers face this all the time.

Most screen readers will parse information from left to write, from top to bottom.[10] Think about a few things when reviewing the order and placement of your words. Is there information critical to performing an action, or making a decision, that appears after (to the right or below) an action item, like in Figure 5.5? If so, consider moving it up in the interface.

Instead, if there's information critical to an action (rules around setting a password, for example, or accepting terms of service before proceeding), place it *before* the text field or action button. Even if it's hidden in a tooltip or info button, it should be presented before a user arrives at a decision point.

Don't Use Colors and Icons Alone

If you are a sighted American user of digital products, there's a pretty good chance that if you see a message in red, you'll interpret it as a warning message or think something's wrong. And if you see a message in green, you'll likely associate that with success. But while colors aid in conveying meaning to this type of user, they don't necessarily mean the same thing to those from other cultures.

10 If you have a working relationship with your developer, it's useful to meet with them and understand how they order the keyboard focus shift. Sometimes, an element at the top, but to the far right might still be the last item in an order.

For example, although red might indicate excitement, or danger in the U.S. (broadly speaking), in other cultures it means something entirely different:

- In China, it represents good luck.
- In some former-Soviet, eastern European countries it's the color strongly associated with Communism.
- In India, it represents purity.

Yellow, which we in the U.S. often use to mean "caution" (because we're borrowing a mental model from traffic lights), might convey another meaning for people in other cultures:

- In Latin America, yellow is associated with death.
- In Eastern and Asian cultures, it's a royal color—sacred and often imperial.

And what about users with color-blindness or low to no vision? And what about screen readers? Intrinsic meaning from the interface color means nothing for them. Be sure to add words that bear context so that if you heard the message being read aloud, you would understand what was being said, as in Figure 5.6.

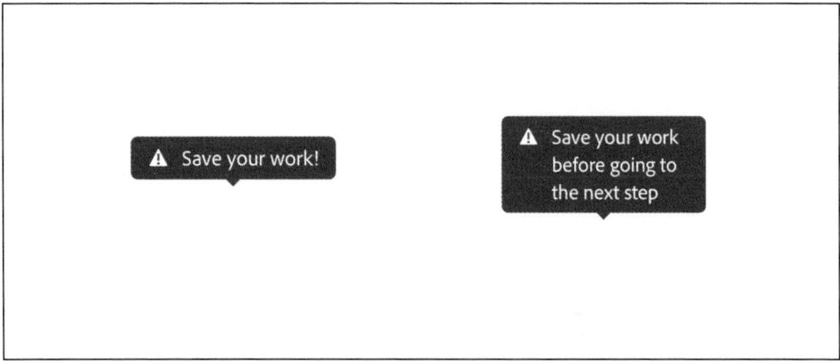

FIGURE 5.6
While a simple in-app message warning a user to save their work before proceeding is more effective, visually, if it is red and has a warning icon, as seen on the left, you should provide more context when possible. The example on the right explicitly says that a user won't be able to proceed to the next step before saving their work.

Describe the Action, Not the Behavior

Touch-first interfaces have been steadily growing and replacing keyboard/mouse interfaces for years, so no longer are users "clicking" a link or a button. But they're not necessarily "tapping" it either, especially if they're using a voice interface or an adaptive device.

Instead of microcopy that includes behavioral actions like:

- Click
- Tap
- Press
- See

Try device-agnostic words that describe the action, irrespective of the interface, like:

- Choose
- Select
- View

There are plenty of exceptions to this rule. If your interface requires a certain action to execute a particular function, and you need to teach the user how their gesture affects the interface ("Pinch to zoom out," for example), then of course you need to describe the behavior. But generally, the copy you're writing will be simpler and more consistent if you stick with the action in the context of the interface itself.

Playstation's Icons Don't Translate

Icons can be the same way. Just look at the classic Playstation controller button layout (see Figure 5.7). Dedicated gamers will know what these buttons usually mean, but it can be obtuse to new gamers on the Playstation platform. Playstation's hardware designer Teiyu Goto explained his choices for the colors and shapes in a 2010 interview with 1UP magazine:

> I gave each symbol a meaning and a color. The triangle refers to viewpoint; I had it represent one's head or direction and made it green. Square refers to a piece of paper; I had it represent menus or documents and made it pink. The circle and X represent "yes" or "no" decision-making, and I made them red and blue respectively. People thought those colors were mixed up, and I had to reinforce to management that that's what I wanted.

In Japan, an X shape is known as *"batsu,"* and bears specific meaning around "no, wrong, go back." And conversely, the circle, *"maru,"* means "yes, continue."

But when Sony introduced the gaming system to Western markets, they reversed the actions. While the "X" can still bear negative connotations (many people draw an X to cross something out, for example), the circle's meaning isn't as inherently clear. The team responsible for internationalization to the West may have thought the "X" marked a target. (A 2019 article on the Verge theorized they could have interpreted the phrase "'X' marks the spot" as supporting this action.[11])

11 Jon Porter, "Japanese PS4s Can Now Use the X Button to Select, but Why Couldn't They Do That Already?" *The Verge*, March 9, 2019, https://www.theverge.com/2019/3/9/18255901/ps4-x-o-cross-circle-remap-firmware-6-50-dualshock-4

FIGURE 5.7
The action buttons on the right side of the Playstation controller have been the same since the Playstation 1's introduction in 1994. But they weren't without their share of international confusion over what the icons on the buttons symbolized.

Pretty soon, the Playstation took off in Western markets, and American and European developers started creating games for it. Those games started being translated for Japan, but kept their Western-centric X-means-yes, O-means-no button actions, and really confused Japanese players.

Finally, in March 2019, Playstation 4 released a firmware update letting players remap their "enter" button preferences. But all this is to say that Playstation built an entire buttons set based entirely on iconography—the shapes on the buttons. They provided very little context around the purpose and function of those icons, and caused significant confusion as the platform grew across the world.

This will fundamentally affect the way you organize your interfaces, and will require cooperation from designers and even product owners.

Finding What's Right for You

It's hard to gauge how much care and intention you put into inclusive design practices for your product, but there's really no situation where your product can't be made better through accessibility or inclusivity, or by taking a longer or deeper look at your users.

If you work in a regulated or government services industry, you and your team are probably required to be inclusive. But, if not, consider looking for accessibility and inclusivity training. If you found the points in this chapter helpful, you'll really get value out of something more formal.

Inclusive, accessible experiences make technology more humane and welcoming. It's not an optional add-in—it should be embedded in the way you work.

CHAPTER 6

Voice

Discovering and Developing Identity

> It was a foggy morning on the streets of San Francisco. You couldn't tell if it was day or night except for the car horns piercing the cool, gray mist every few seconds. The private eye sat in his fourth-floor office, finishing his usual breakfast of cigarettes and coffee. The phone rang. The private eye took his wingtips off the desk, leaned forward, and answered the phone:
>
> "Hey diddly ho there, friend!"

Notice how jarring that line of dialog was there. You were expecting some hard-bitten Dashiell Hammett[1]-type who's cynical but with a heart of gold, and instead you got your perpetually upbeat, cheeseball neighbor. Unless this was a novel about how Ned Flanders from *The Simpsons* was San Francisco's newest gumshoe, the writer wasn't doing a great job of matching the voice to the context of the book.

But it doesn't need to be the same for you! Your product's voice helps you connect with your users. It helps set expectations, motivates them through an interaction, and directly ties a brand directly into the products it makes.

It's also a useful tool to align a team of writers to produce consistent, well-considered content. An effective voice strategy lends tangibility to the abstract entity of a company, or a product, or some other work produced by people, and lets the writer make specific decisions about style, grammar, pronouns, and more.

Finding a Product's Voice

Developing and maintaining a voice is a constant, ongoing effort. One person who has done a great job with this is Anna Pickard, Head of Brand Communications at Slack, a collaboration and messaging platform used in many workplaces.

"I was brought in to oversee the writers at the company. Although, arguably, it was a company of writers—everyone here is very good with words. Our founder was a Philosophy PhD candidate, others were English students, et cetera."

1 A prominent detective fiction writer from the early 1900s, whose work influenced Hollywood's film noir movies.

Her job seemed to be less about doing the writing, and more about documenting and steering it.

Pickard started distilling what Slack's voice really was. To do that, she audited all the "voicey" places where a strategic style of writing was important, and she realized they were all over the place at the company.

"The voice, as it was originally, showed up in places like Slackbot's initial scripts, like 'Welcome to Slack" and 'I'm Slackbot, I'm a little dumb, but I'll try my best.'" (See Figure 6.1.)

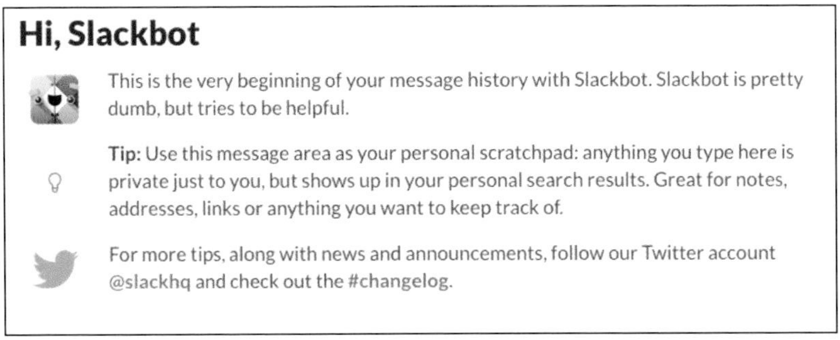

Hi, Slackbot

This is the very beginning of your message history with Slackbot. Slackbot is pretty dumb, but tries to be helpful.

Tip: Use this message area as your personal scratchpad: anything you type here is private just to you, but shows up in your personal search results. Great for notes, addresses, links or anything you want to keep track of.

For more tips, along with news and announcements, follow our Twitter account @slackhq and check out the #changelog.

FIGURE 6.1
One of the first on-voice parts of Slack in 2013: Slackbot. It's an automated, meta-messaging area in Slack, sort of a combination of onboarding tour guide, task reminder, and account maintenance guide.

It became clear to Anna that "voice" wasn't just about a persona that a product channeled. It was about style, word choice, structure, and an editorial point of view. And while Slack's product sits at the forefront of their company, it manifests all throughout the brand. See Figure 6.2 for how the Slackbot welcome message looked in 2019.

In this chapter, we'll dig into the concept of "voice," what it means strategically, and how to document it and implement it tactically in your digital experience.

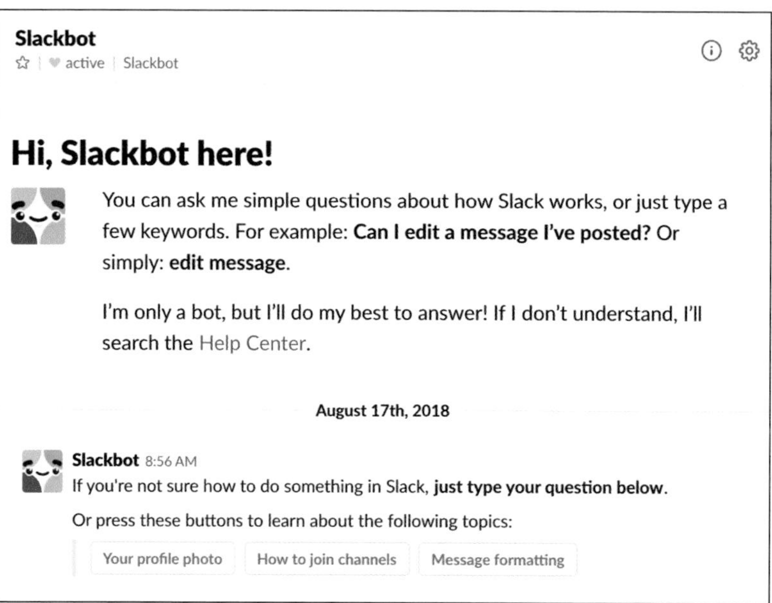

FIGURE 6.2
How Slackbot manifests today, in April 2019.

Brand Voice vs. Product Voice

There is a lot of confusion around brand voice and product voice. And rightly so: they can be the same, or they can be very different. Some companies have a voice that's primarily product-led. And some give their products a brand-driven voice. We'll dig into the differences between them.

A *brand* voice is the consistent, goal-driven expression of a brand through words. It serves to engage an audience or to motivate them to become customers. It's how a brand would speak and write if it were a person. In essence, it's an expression of the brand's personality, and should be expressed in everything that the company produces—brochures, customer service phone trees, business cards, websites, social media presences, apps—everything. Often, even employees will receive coaching on how to speak and write in accordance with the "brand voice."

A *product* voice is a bit more tactical than that. It definitely needs to fit within and carry the brand voice, but it might have a narrower set of goals and face a stricter set of constraints. It's there to define what an app, a website, or a digital product produced by this brand sounds like.

Sometimes they're the same thing. If you're writing for a small organization that focuses on making one app (like Slack was when Anna began), the two are so intertwined there's no real difference between them. If you're at a big company that has multiple products for many audiences, or if the product is only a small part of what they do, there might be clear boundaries.

Take Adobe, for example. It's a large, mature tech company that makes digital products for creatives, office workers, and marketers. In other words, it encompasses a wide swath of people with vastly different experiences and goals. How does Adobe's brand voice resonate with all these people, but still sound like one, cohesive brand?

Their brand strategy team has a core set of principles they follow in order to do this. They aim to be:

- Captivating
- Stimulating
- Fresh
- Proactive
- Approachable
- Expressive

This is all fantastic work and gives a writer or team of writers the ability to scope their writing, but when faced with writing a small piece of microcopy, you don't have the room to be "captivating," "stimulating," and fit into these other characteristics. More often than not, their main goals are to:

- Be clear about what action a user is taking
- Make sure that the user understands the concept of what they're interacting with
- Balance the right amount of explanation with brevity so that the user isn't cognitively overloaded

Of course, the brand voice aims to accomplish these goals, too, but ultimately, it's there to serve the brand, not the user's experience in a product. The handy thing about developing a voice strategy, be it for a brand or product, is that the methodology is the same.

KEEPING VOICE ACTIVITIES SMALL

If you work for a larger company or even one where it's a bit of a challenge to get things approved, creating lightweight, informal product voice guidelines for your team can be helpful. Rather than calling in the CEO and all the marketing stakeholders, you can work with your immediate team to find a voice that works.

As long as it's connected to or inspired by the larger brand voice, it can be a great way to give writers the resources they need to keep things consistent without having to wade through lots of meetings and approvals.

I've done this in the past with a few teams, and we usually just use a shared document or wiki page to keep track of it, along with examples. ■

Voice Attributes

Before you can work your way to a set of rules guiding how you or a team of writers choose how to write, you have to figure out what you want that voice to sound like. Do you want to come across as simple and powerful, like Apple (see Figure 6.3)? Do you want to be refreshingly clear and compelling like Capital One? Do you want to show that you're at the intersection of art and science, like Adobe? This is where branding exercises will help, if that work hasn't been done already, but you can also just develop principles on your own that you can use to explain your work.

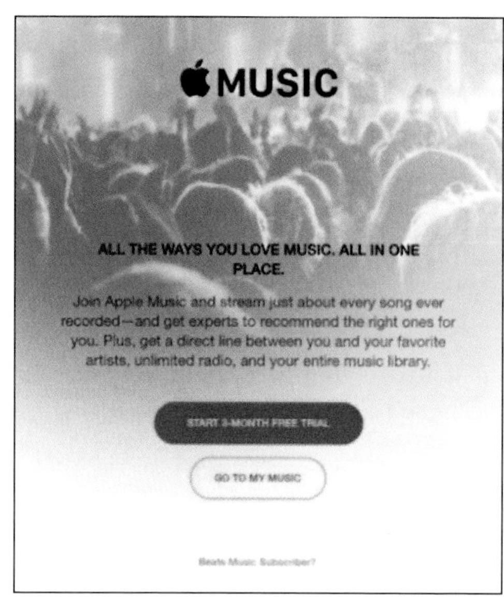

FIGURE 6.3
Apple's voice is powerful, but simple, as seen in this Apple Music onboarding screen.

Airbnb's talented content strategy team has spent a lot of time on their voice principles. At the time of this book's writing, they had declared their product voice to be:

- Straightforward
- Inclusive
- Thoughtful
- Spirited

Think about the last time you used Airbnb (if you have at all). Perhaps you noticed these principles woven throughout the interface writing, as shown in Figure 6.4.

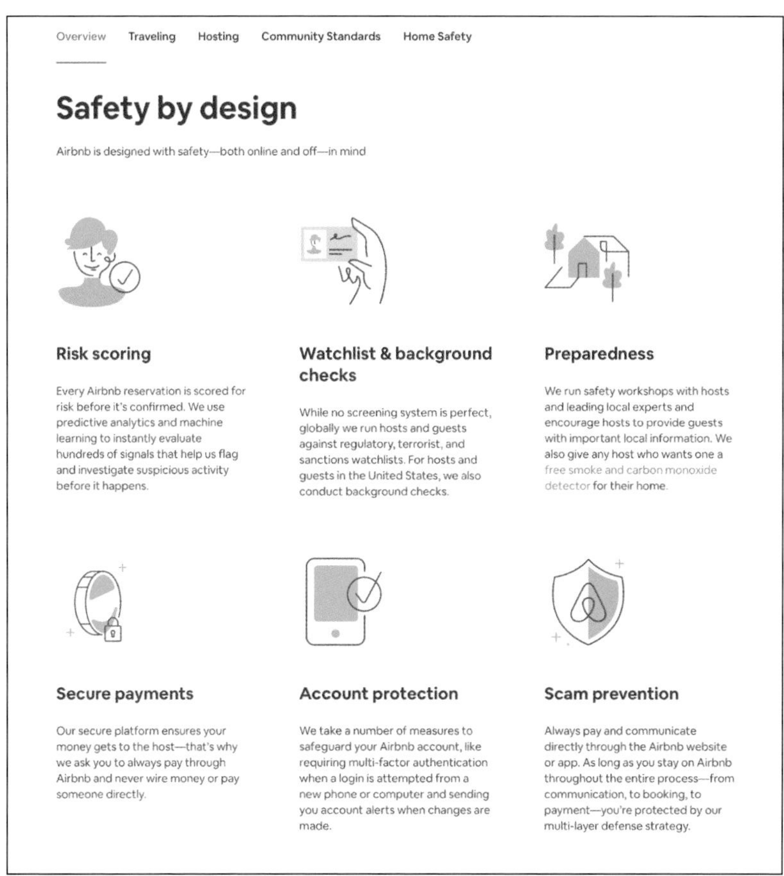

FIGURE 6.4
A page about Airbnb Trust and Safety, which embodies their voice principles: straightforward, inclusive, thoughtful, and spirited.

The output of these principles seems simple and straightforward, but each of those adjectives were likely painstakingly discussed, tested, explored with a thesaurus, pitched, and ultimately approved for use. And while Airbnb has chosen not to share the documentation and examples of those voices in action, next time you use Airbnb, apply an editor's eye to the interface, and you'll probably see those principles reflected in the writing.

There are several different proven ways to really document and express voice principles—we'll look at just a couple of them here.

This, but Not *That*

In their book *Nicely Said*, Nicole Fenton and Kate Kiefer-Lee talked about a method of developing a brand voice called *This But Not That*. They said:

> It's easy: List some words that describe your brand, and then explain each one by what it *doesn't* mean. That second word helps writers better understand each personality trait.

They used voice documentation for Mailchimp, a well-known email newsletter and marketing software company, known for its cheerful personality and monkey mascot, Freddie. Here's Mailchimp's voice, according to Fenton and Kiefer-Lee:

Mailchimp is . . .

- Fun but not childish
- Clever but not silly
- Confident but not cocky
- Smart but not stodgy
- Cool but not alienating
- Informal but not sloppy
- Helpful but not overbearing
- Expert but not bossy
- Weird but not inappropriate

This is an interesting way of documenting voice attributes—the first word sets a baseline, and the second word (which isn't an opposite work, but a version of the first word taken to the extreme) sets a hard outer limit.

Declarative Statements

In late 2018, Mailchimp announced a major rebrand. In addition to a new logo and a more "grown-up" brand typography system, they refreshed their voice principles.

This time around, they decided to use simple, declarative statements, with some short rationale to add flavor and nuance. This new direction gave the team a North Star—guiding principles they could aim for. This is something you and your team could use as a point of conversation when you start talking about how voice should be applied to your writing.

At the time of this writing, here's what they said:[2]

. . . (W)hen we write copy:

1. **We are plainspoken.** We understand the world our customers are living in: one muddled by hyperbolic language, upsells, and overpromises. We strip all that away and value clarity above all. Because businesses come to Mailchimp to get to work, we avoid distractions like fluffy metaphors and cheap plays to emotion.
2. **We are genuine.** We get small businesses because we were one not too long ago. That means we relate to customers' challenges and passions and speak to them in a familiar, warm, and accessible way.
3. **We are translators.** Only experts can make what's difficult look easy, and it's our job to demystify B2B-speak and actually educate.
4. **Our humor is dry.** Our sense of humor is straight-faced, subtle, and a touch eccentric. We're weird but not inappropriate, smart but not snobbish. We prefer winking to shouting. We're never condescending or exclusive—we always bring our customers in on the joke.

2 "Voice and Tone," *Mailchimp Content Style Guide*, September 2019, https://styleguide.mailchimp.com/voice-and-tone/

You can see the evolution of the documentation—"this but not that" principles like "smart, but not snobbish," and "weird but not inappropriate" were wrapped together to give extra description to their statement about humor.

These principles are great because they really bridge what can often be a gap between business goals and brand creativity. Mailchimp has built a reputation on being really fun and delightful to use—that delight was infused through the design, the copy, and the workflow—but (and this is just speculation) there's a good chance that big, enterprise companies thought that it was too unprofessional and signed up with one of their more serious competitors instead. This documentation shows their team of writers (and those nerds like us who love reading about product voice principles) where that line is: they're allowed to be subtly humorous, but not at the expense of clarity, education, and authenticity. And you might notice that they mention their customers in all four principles, so it shows their dedication to always keeping the customer in mind when writing.

Practical Tips for Principles

You can use these principles to dig even deeper and make some style decisions to really align writers around a point of view. Let's say that your product has a "human" or "conversational, not idiomatic" voice principle. You want to find practical ways to sound like a person, but not to such an extreme where it gets culturally specific. How do people talk in conversations? Well, for starters, they do the following:

- Vary sentence length and structure for more natural readability
- Use contractions like "isn't" and "you'll"
- Always read their writing aloud to make sure it sounds natural and not stilted

If you're in charge of maintaining or contributing to a larger voice documentation, you'll start to notice patterns among successful messages that you test (more on that in the next chapter about tone). Start to look at microcopy and messaging opportunities that seem off-brand, or in a voice that doesn't seem to fit with the larger system. Discuss among your team, or if it's just you, loop in a few product owners and designers:

- What about it seems off-voice?
- How do you change it to fit in better?

And, of course, document your discussion and add it to the rest of your voice documentation. A good voice system is always evolving.

FOCUS ON WHAT'S IMPORTANT

There are lots of great examples of voice guidelines out there, published on the web for the world to see. However, keep in mind that guidelines aren't nearly as important as how the voice is used in whatever work you're building.

In my career, I've never worked in an organization that has had polished, public-facing voice guidelines, but voice has been an important part of how I've written and designed for the products I've worked on. I've socialized voice ideas by:

- Influencing product vision and planning
- Including voice rationale in design deliverables
- Speaking to voice decisions as I present my work

What's important is that you know what you're trying to accomplish, not that you have great guidelines. Guidelines that don't affect the end product aren't worth much. ■

Voice Principles for Product Experiences

There's a common thread that weaves throughout most of the voice principles for digital experiences. Capturing the essence of the brand is important, but it should all be in service of three key principles that underlie the importance of writing the user experience. Here are three principles we've observed in the industry and use in our own work, in order of importance: be clear, concise, and human.

Consider how these attributes fit into the vision for your product when developing voice principles.

Clear

We wrote about this in Chapter 3, "Creating Clarity: Know What You're Designing," but there's nothing more frustrating to a user than an interface that's unclear. It's critical to break down concepts, ideas, and actions into digestible, understandable chunks. Brand voice decisions should never sacrifice clarity.

Concise

You have limited time and space to communicate with users of your product. Unless it hinders clarity, it's important to be as simple and to-the-point as possible. In *The Elements of Style,* a foundational book for writers, Strunk and White say to "omit needless words." We'd take that a step further, and instead of focusing on the number of words, focus on distilling the concepts, terms, actions, and ideas you're presenting to your users. Perhaps a better rule is "to omit needless information."

Human

After clarity and concision, writing with a natural rhythm, conversational flow, and with empathy will reduce friction for readers and users of interfaces. We're not saying you should use slang or pack your writing with idioms—remember, inclusive writing is the goal, and slang limits the reach of your product to those with insider knowledge of what it means.

Pickard talked about how these three universal principles—which Slack actually uses as their voice principles—can tie directly to their mission statement of "Make work life simpler, more pleasant and more productive."

"It's funny," she said, "these principles come out of trying to work out the right way to express to people over and over again that this is the way our voice should be. I've done it by using our mission statement [seen in Figure 6.5]: *'Clear'* makes work *simpler. 'Concise'* makes work *productive.* And *'human'* makes work *pleasant.*"

Make work life simpler, more pleasant and more productive.
Slack is the collaboration hub that brings the right people, information, and tools together to get work done. From Fortune 100 companies to corner markets, millions of people around the world use Slack to connect their teams, unify their systems, and drive their business forward.

FIGURE 6.5
Slack's mission statement. Pickard can tie the brand's voice principles directly to these three points in their mission statement, seamlessly connecting the voice to their strategy and values.

Let's see how these elements interplay in a few examples from other apps and digital experiences.

Bitmoji App Update Message

Bitmoji is a product that gives users a way to create avatars of themselves that they can insert into things like text messages or social media posts. Despite the comic strip–like format of this message, a user might see the message in Figure 6.6 when they open the Bitmoji keyboard on their smartphone, when the Bitmoji app needs to update.

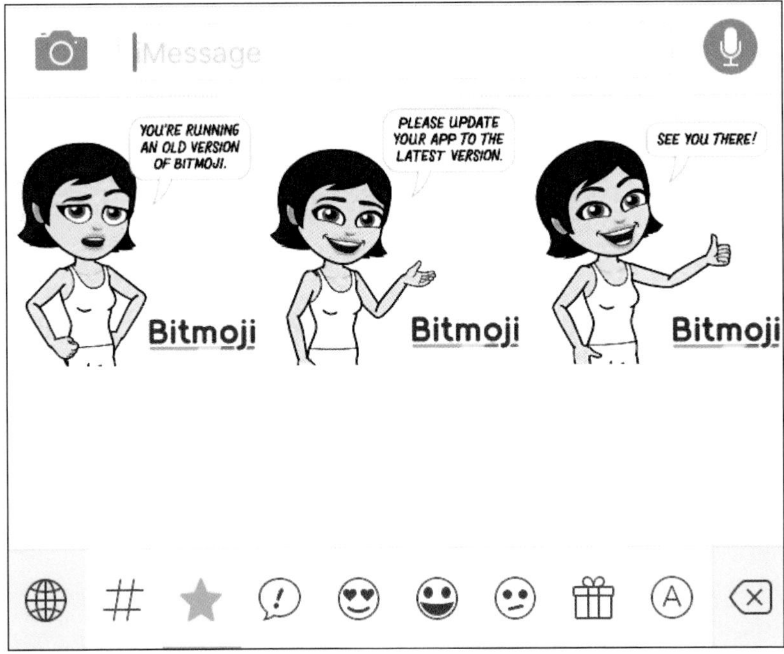

FIGURE 6.6

Bitmoji, an app that lets you share little cartoon avatars of yourself, uses your avatar to tell you to update the app.

How does this message fit with our three principles?

Clear: The message is straightforward—following our error state methodology from Chapter 3, it's explaining to the user what the problem is and how to resolve it.

Concise: It breaks the idea into simple chunks and doesn't use a lot of extra words or give unnecessary context.

Human: All the while, it sounds like a person is saying it to the user, and gives the feeling of friendliness. The "See you there!" is perhaps unnecessary, but adds no extra cognitive load and really finishes it out nicely. Bonus points for the illustration—what could be more human than a version of *you*, the user, giving *you* the message!

Facebook's Post Prompt

Facebook's prompt to help users put together a post is clear, concise and human as well (see Figure 6.7). Over the years, they've experimented with different kinds of prompts, like "What's your favorite color?" or "What did you have for breakfast today?" but then users would end up with posts with no context like "Purple" or "Waffles," so they have always gone back to the age-old prompt of "What's on your mind?"

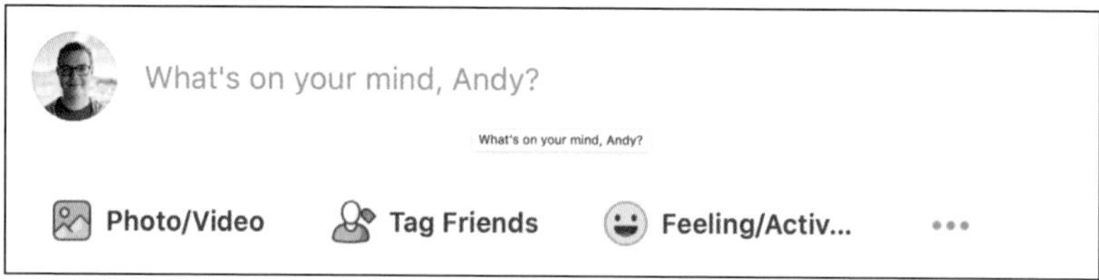

FIGURE 6.7
The ubiquitous composer prompt at the top of Facebook's News Feed. Although it's taken on many forms and features over the last decade, this prompt has remained steady for years.

How does this work with our three principles?

Clear: In just five words, including your given name, it's easy to see what you're supposed to put here.

Concise: This message doesn't muck around with extra information—there's no extra message like "Just type what you're thinking, and we'll share it with your friends!" That's completely unnecessary.

Human: This prompt uses contractions ("what's"), builds some connection by calling you by name, and although it's a little idiomatic ("What's on your mind" wouldn't translate literally, it's asking you for your thoughts on something), it's universal and timeless enough to get the gist through to most English speakers.

Your Voice Should Always Be Evolving

If you have a fun, quirky, modern brand, and you're building a product that can afford to also be fun, quirky, and modern, it makes sense for its brand voice and product voice to be in parity. But as your business grows, and your user base grows with it, you'll start to notice that messages that rely heavily on this type of voice lose their impact at best, and confuse people at worst.

Look at Lyft. Founded in 2012, it's a rideshare company positioned as the friendly, quirky alternative to the more business-like and aggressive Uber. If you've ever been in a city where Lyft and Uber are competing side-by-side, you may have used both. The functionality's the same, but in Lyft's early days, the app's interface elements were bright pink, featured fun illustrations, and once your ride came (sporting a fuzzy pink mustache on the car's grill, of course), you were encouraged to sit in the front and give your driver a fist bump.

And if you were a new passenger with Lyft back in 2014, you might have gotten an email like Figure 6.8.

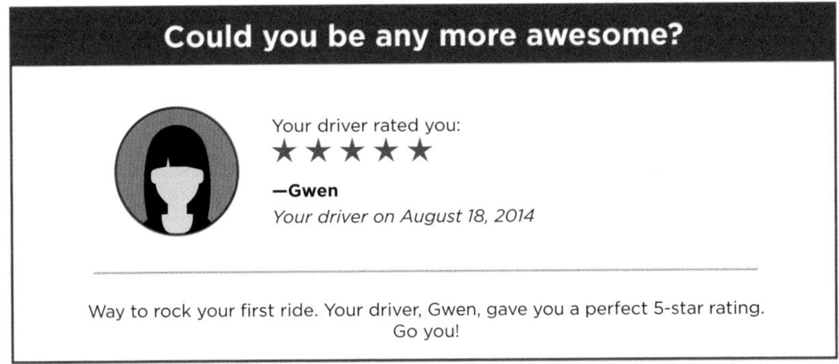

FIGURE 6.8

A celebratory email from Lyft in 2014 congratulating a rider on getting a perfect star rating from their driver.

It seems perfectly on-brand for them. Cheerful, celebratory, and conversational. It's just the way someone driving a mustachioed car might congratulate you on being a good passenger. That headline almost channels Matthew Perry's character, Chandler Bing, from the 1990s American sitcom *Friends*—and anyone familiar with the show might put the emphasis on the "be" in that rhetorical question—"Could you *be* any more awesome?"

But as Lyft grew, and a more and more diverse set of passengers started to ride, that phrase lost its meaning. A Chinese coworker, fluent in English but lacking the cultural context to process that rhetorical question as Lyft intended, was puzzled: "If I got a perfect 5-star rating, why does Lyft want me to be more awesome?"

That certainly had the opposite effect of the celebratory message Lyft was trying to get across. As the company grew (and at the time of this writing, continues to grow!), they tested, iterated, and optimized their language to scale with it.

While we don't have a comparable email from Lyft recently to compare it to, this end-of-2018 summary email in Figure 6.9 shows that an equally celebratory message isn't quite as idiomatic and culturally specific. "Cue the music and toss the confetti" is translatable and easy to parse, even if confetti isn't necessarily part of your cultural context.

FIGURE 6.9

An email from January 2019 from Lyft, summarizing the previous year's activity. It's celebratory, cheerful, and conversational, but not idiomatic and culturally specific to the United States.

Scaling Your Voice

Hopefully, if your user base is growing, your writing team is, too. That's not always the case, of course—your company may not value (or know the value) of strategic words and shared language.

If you're in a position of leadership to align and scale the team, great! Here are some tips and things to keep in mind to make sure that you're scaling your voice thoughtfully.

Documentation is key, but be flexible. Be sure that you're keeping good records of what you write.

Pickard suggests:

> List out every "we do this, we don't do that" rule. Like for our Twitter presence, we say:
>
> "We don't use jargon, we don't 'lol,' we don't culturally appropriate, we are elegant with language, we're well-read, not showoffs, and somewhat esoteric. And yet, every single one can be caveated with ...unless it's appropriate or funny."

Too few rules leave the voice too abstract for writers to align on. Too many rules leave them paralyzed and unable to work out how they should sound.

Get Feedback on Your Work

If you have a community of writers or communicators from across a range of experiences and contexts, try to get them in a room and show them your work. We'd even recommend getting that feedback from people who are outside the product work, from different roles.

Read Aloud

It sounds silly, but one of the best ways to get a sense of your writing is to hear it being read aloud, even if that reader is yourself. Just as a musician will hear an off-tune chord or a missed beat when they play the music they write, you can hear if your cadence or pacing makes sense, and if it's "on-voice" or not.

IN PRACTICE

Establishing Voice Principles

Let's say that you're on a team creating a peer-to-peer payment app for a big, national bank: ABC Bank. That bank's brand voice is present everywhere it interacts with people—from phone book advertising all the way through the thank-you message on ATMs. The tellers are even instructed on how to speak in the ABC voice.

Those voice principles call upon the brand to be:

- Friendly
- Knowledgeable
- Reliable

But here's an interesting complication: your app is an oddity in ABC Bank's digital product line. It's not there for the bank to talk to you and give you information like account balances, loan information, branch finders, and other typical banking app features. It's there to let you interact with your friends—you can send money, request and accept money owed to you, and "chat" among your friend networks.

Some questions you should ask yourself:

- Do users expect the bank's brand to be carried through into the interface? Or would the brand fade into the distance to allow their own voices to shine?
- Our targeted demographic is younger than the typical customers of the bank. Do we flex our voice to appeal more to younger customers, or stay on brand?

After you've thought through some of these considerations (often with executives, product managers, product marketing folks, and others who think about business goals), it's time to start developing your voice principles.

Leadership has asked you to use the bank's brand voice as the basis for your app. So let's explore the stress-case limits of each of those characteristics to make sure that you and your writing team aren't over-relying on that characteristic.

After much discussion, you come up with:

- Friendly, not familiar
- Knowledgeable, not pedantic
- Steady, not boring

Great! This will help people understand how an extreme example of these principles can harm your brand and turn off users. Now, add a little justification for why you made these choices:

Friendly, Not familiar

- Show warmth, encouragement, and motivate new users to try your features
- Don't use idioms or trendy language to sound cool or hip

Knowledgeable, Not pedantic

- Help guide users through potentially sensitive flows involving payment information and money issues
- Identify places where they might need a bit more explanation, but don't overload them with unnecessary information

Steady, Not boring

- Build trust—customers are using this app because they know and expect ABC Bank—with over 100 years of experience—to handle their money safely
- While we have a heritage of responsible financial handling, we don't want to be out-of-touch, so customers should expect a modern, pleasant digital experience

When Voice Takes a Back Seat

As we've discussed, voice is great for connecting a product, an experience, or an interface to a brand. It really sets expectations for a user as to what kind of a company is behind the thing they're using. A strong display of the brand through its voice is great for collaboration apps, games, and instances where you're trying to build a connection between users and the company behind the product. The voice is an obvious, central part of the interaction in these cases.

But some interfaces shouldn't put that kind of focus on the brand. Their goal is to get someone through an interaction. And with some apps, the people who use it are in a hurry, stressed out, or perhaps even impaired.

Let's look at the app for Geico, the car and home insurance company, for example. If you're familiar with the company, Geico has a strong brand voice. Featured in many, many television commercials over the last decade or so, the Geico gecko, with his British accent and innocent but curious demeanor, is inextricable from the brand he represents. It's a company with a strong brand voice.

In their app, users can pay their bill, sign up for new insurance, look up their account number, and change mailing addresses. And that same friendly gecko-like voice is present throughout.

But what about reporting an accident or getting roadside assistance? As soon as you dig into that section of the app (as seen in Figure 6.10), the voice shifts to being more supportive and goal-oriented. Suddenly, the app is like an emergency dispatcher, trying to get the information it needs from you in order to send help. The friendly phrases and exclamation marks drop away, and it's as simple as it can be.

(This starts to get at what you'll learn in Chapter 7, "Tone: Meeting People Where They Are,"—how you can shift the tone and structure of your writing in different contexts to meet users where they are.)

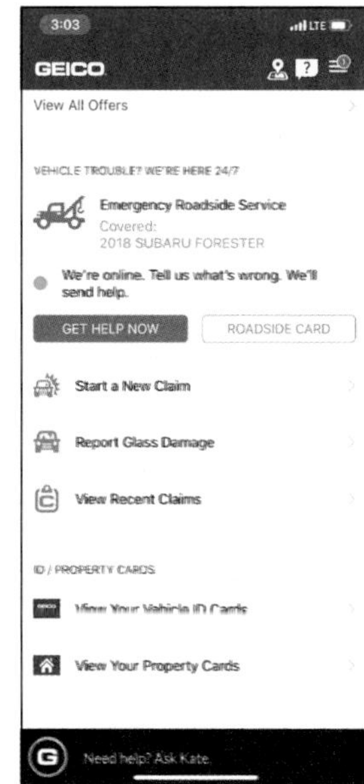

FIGURE 6.10

The loading screen of the Geico insurance app brings in a lot of the personality of the brand, but in urgent situations, the brand recedes into the background.

DEVELOPING A VOICE FOR THIS BOOK

We've worked together a lot—building and evolving a UX writing workshop that we've presented several times at multiple conferences. One of the reasons we work so well together is because our distinct voices and personalities complement each other. Michael is deeply thoughtful and thinks a lot about the big picture. Andy leans more toward the details and the process.

This works great when we're on stage or in front of a classroom together, but what about this book? How do we combine our distinctly different voices into one? As one reviewer of our book proposal mentioned in their feedback, "This book is (in part) about voice: you're really going to have to nail it."

We decided to eat our own dogfood, as the saying goes, and develop some voice principles for our book!

This book's voice will be:

- Instructional, but not dumbed-down
- Conversational, but not overly familiar
- Confident, but not a know-it-all

- Passionate, but not theatrical
- Pragmatic, but not prescriptive
- Entertaining, but not goofy

As we wrote the book, we kept each other accountable by editing each other's work, and even getting on an all-day video call to discuss our decisions.

How did we do? ■

Finding What's Right for You

It's important to make sure that your product's voice isn't getting in the way of your users' goals. You'll want to ask yourself and your team some questions to balance branding with ease of use:

- **Speed vs. Engagement:** How important is it to get the user through an interaction?
- **State of Mind:** What is your user likely to be thinking about and feeling as they use your interface?
- **Brand Moment vs. Utility:** How important is it to build brand awareness at this point of the user's journey with your product?

And remember, "voice" doesn't necessarily mean that you have to pick a unique way to write with a strong personality. Sometimes, choosing a "voice" means that your product is speaking in the plainest, simplest way possible.

CHAPTER 7

Tone

Meeting People Where They Are

As a child, you might realize that the way you talk to your parents, for example, is different than the way you talk to your friends. And later, you realize that you talk to your grandparents in a different way than you talk to your parents.

It just gets more complicated as you get older. As you get better at communicating with others, you might start changing your tone and word choices depending on whom you're talking to, such as:

- Friends your age
- Friends significantly younger or older than you
- Work friends
- Coworkers you don't know well
- Coworkers who are in other countries than you
- Or a myriad of other contexts

For many people, this context-switching and tone variation comes naturally. But it doesn't come naturally to software, or even to software-building teams. That's why you need a strategy for switching tone when appropriate.

LANGUAGE IS NOT STATIC AND ABSOLUTE

When I was a kid, I was extremely pedantic and particular about the language I used. I hated using slang, or at least what I defined as "cool" slang—the words other kids in the early 90s were using. In an effort to find order and precision in my youth, and because I was already discovering a joy and penchant for words, I leaned toward literal, descriptive words.

"No, Kelly," I'd say, correcting my little sister, "That's not 'cool,' that's 'interesting.'"

(And yes, I know that I was an insufferable child.)

As I got older and gained self-awareness and context for my words, my tone softened. I realized that when I was talking to my friends, I needed to adapt my parlance to include more words like "cool" and "awesome" and "wassup" (it was the 90s after all), but when I talked to my grandparents, those words didn't fly. ■

What's the Difference Between Voice and Tone?

These two terms, voice and tone, are often conflated, even by those who are writing. You pair the terms together so often that "voice and tone" become one. Some people even use "tone of voice."

In a nutshell (and as we showed in the last chapter), *voice* is the personality that your brand, or product, or digital interface manifests, which sets it apart from others. *Tone* is the way that voice is expressed in certain contexts, i.e., to respond to or guide a user through a particular workflow or interaction.

Tone changes might manifest in different ways:

- **Word or phrase choice:** Are you using short, concise words and phrases, or do you take the time to add in pleasantries or explanations?

  Example: "404 Error: Try Again" vs. "We couldn't find the page you're looking for."

- **The structure of the message:** Do you lead with the benefits of a feature or an action? Or are you just describing what is happening on the screen at this very moment?

  Example: "Reset your password now" vs. "Keep your account secure by resetting your password."

A Channel Switcher, Not a Volume Knob

Traditionally, copywriters and communication strategists—usually in a marketing-driven organization—think of "tone" as similar to a volume knob that you turn up and down, choosing how much style or brand voice you infuse into your writing. Many brand and communication style guides map different types of content to different levels of tone infusion, as seen in Figure 7.1. This could include high-visibility, high-impact things like headlines, marquees, and email subject lines, to small, utility copy-like button calls to action.

Why Is Tone So Important?

Someone who has spent a lot of time studying tone is Dr. Melanie Polkosky, an expert designer and researcher who specializes in how language is used in software. She has a great deal of experience designing voice interfaces, but also applies her skills to web and mobile apps.

During her career, she wanted to determine which aspects of an interface were most important for usability (the ease of access or the use of a product or website), so she conducted a large study using a technique called *factor analysis* to figure it out.

In particular, her research highlighted the importance of finding the appropriate tone an interface should speak in. The study included 862 participants and evaluated 76 items that could affect usability. Her study found that customer service behavior was one of the most important categories, and it included many tone considerations to prioritize when writing. Here's what her study said:[1]

> **Customer service behavior included items that were related to the friendliness and politeness of the system, its speaking pace, and its use of familiar terms.**

1 Melanie D. Polkosky, "Toward a Social-Cognitive Psychology of Speech Technology: Affective Responses to Speech-Based E-Service " (PhD diss., University of South Florida, 2005).

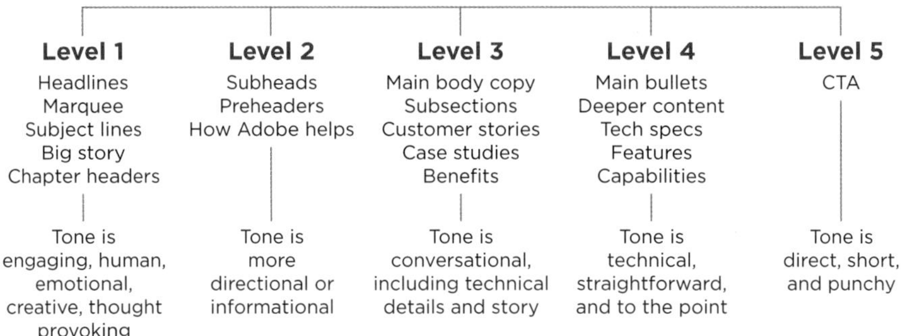

FIGURE 7.1

An example of a tone hierarchy from a large software company's marketing and communications style guide.

This discovery helped Polkosky make the case that the right tone plays a role in the usability of a digital product.

"Tone is so incredibly important because I think it's the thing that kills most interfaces," she said. "You know pretty immediately when you're talking to someone if they're a jerk, or if they think they're better than you, or if they seem pretty friendly, or if they're talking to you in a way that seems respectful. The same social skills govern our perception of technology, as well as humans, and determine whether we're willing to engage in or persist with an interaction."

She has used this insight to help her teams prioritize tone, giving her data she can point to. "If you're going to argue with me about the importance of tone, I'm going to come back at you and go 'here is an empirical study that has held up for almost 15 years now' to show that out of all these hundreds of things that could be important, tone is one of the top four," she said.

Polkosky recommended always taking research when reviewing your work with members of the team, even if it's as simple as referencing a study, a book, or an informal test with people in your neighborhood. "Take it out of the realm of my opinion versus your opinion," she said. "Bring your opinion, but say here's the data behind my point of view."

For Polkosky, her users are her motivation. "Human communication is the most important gift we have, whether it's through speech or writing," she said. "I really do think that human beings are worth fighting for, especially as technology overtakes more and more of what it means to be human."

If you remember from the previous chapter, Adobe's brand voice strives to be the following:

- Captivating
- Stimulating
- Fresh
- Proactive
- Approachable
- Expressive

The ultimate expression of these tenets is captured in Adobe's "highest volume" level, and they diminish in intensity from there. It makes sense, really for marketing communication. It's predominantly one-way—it's a megaphone through which writers are sending their message to listeners through email, website copy, and sometimes even ancient technologies like billboards and printed material.

But digital is different. You're not always communicating in such a directional way. Writers don't have the luxury of always setting the topic and crafting a message around it. Communication doesn't flow so linearly; it's contextual, and things change depending on the following information:

- Where users are in the journey
- How experienced they are in using the interface
- What their intentions are
- What their mood is: how receptive they are to what you are guiding them to do.

You can ask a series of questions, developed by the Facebook Content Strategy team, to figure out how to build different tones for your product, as shown in Figure 7.2.

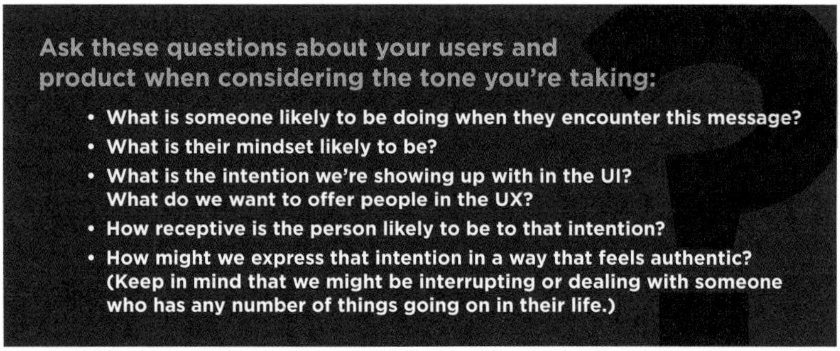

Ask these questions about your users and product when considering the tone you're taking:

- **What is someone likely to be doing when they encounter this message?**
- **What is their mindset likely to be?**
- **What is the intention we're showing up with in the UI? What do we want to offer people in the UX?**
- **How receptive is the person likely to be to that intention?**
- **How might we express that intention in a way that feels authentic? (Keep in mind that we might be interrupting or dealing with someone who has any number of things going on in their life.)**

FIGURE 7.2
A list of questions to consider when building your tone framework, as developed by the Facebook Content Strategy team.

Because there's added dimension to the context you want to use while communicating, you tend to think about tone as a spectrum upon which you pick an intended tone, similar to Figure 7.3.

Writing for digital experiences is a conversation; however, it's not necessarily a conversation in the traditional sense where two or more parties are verbally talking back and forth (although with the rise of chatbots and voice interfaces, that's becoming more and more relevant). Rather, it's conversational in the sense that it flows with a give-and-take of information, a rhythm, and a contextual understanding of the needs of the user. You can often look to real-life conversations as inspiration. There are two main factors to consider, which are the two parties in a conversation: you (the interface) and your users.

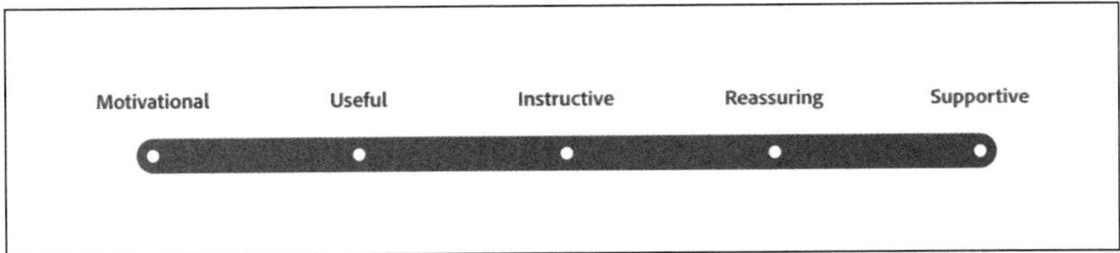

FIGURE 7.3
A spectrum of tone from Adobe.

A Robust Tone Framework at Scale

When new content strategists start at Facebook, they attend training sessions in how to choose the right tone to use when they are writing. One of the things leaders really drive home is that in UX writing, it really comes down to how to best mirror how you might interact with someone in real life.

For example, if you sat down with your best friend and started to tell them your problems, they might lean in to show that they're actively listening to you. They might give you nonverbal cues like nodding their head, or grimacing when you tell them about something awkward or negative. They might nod their head and offer encouragement for you to go on. Or they might mirror your body language—cross their arms, perhaps, or tilt their head, to make you comfortable by expressing themselves in the way you express yourself.

This mirroring is a subtle way that your friend is demonstrating trustworthiness to make sure that you feel comfortable opening up to them, and to show you that they're empathetic toward what you're going through.

And while the software you're designing doesn't have the same sort of relationship with its users (nor should it!), there are ways you can demonstrate your trustworthiness and help them get through a particularly difficult or intense interaction.

Alicia Dougherty-Wold is VP of Product Content Strategy and Design at Facebook, and leads one of the largest product content strategy groups in the world, supporting Facebook, Instagram, Messenger, WhatsApp, Oculus, and other apps and products.

Jasmine Probst is a Director of Content Strategy whose work has also spanned Facebook's family of apps. Together, Alicia and Jasmine have worked to help teams use tone consistently across a broad range of product experiences for more than two billion people.

"In the early days of Facebook, and even in the middle days, the whole ethos of Facebook's design was to recede into the background and let people's content come to the foreground," Dougherty-Wold said. Users could potentially see anything from birth announcements to natural disasters in their News Feed.

"We were starting to see contexts where we needed to humanize how Facebook communicated," she said. "We needed to meet people where they were, with different emotions they were having."

This "meeting" could range from the emotional gamut of happiness (for birthday or anniversary announcements) to empathy or sympathy (if a friend passed away and named you as their legacy contact). That's a lot of responsibility, and one that Dougherty-Wold and her team didn't take lightly.

There was a really good organizational reason for making a framework, too—the team was growing.

"We went from a very small team where we could keep voice and tone consistent by having conversations, showing work to each other, or having content critiques," Dougherty-Wold said. "That works when you have 10 or 20 people on a team. But when you have dozens, and you need to be communicating consistently in a way to take care of a really diverse global audience, you need tools to do this, and do it well."

Probst explained: "It gives us a way to show in more concrete terms what it is that content strategists do, and think about, to partners."

Probst pointed out that "tone" is often viewed as solely about craft. However, she said, a framework for tone grounds that craft in a strategic approach, so that a team can reproduce it more consistently and objectively.

"Humans, when they interact with each other, want to modulate intensity, modulate (physical) closeness, and communicate empathy very intuitively," Dougherty-Wold continued. "We never want to presume how someone is feeling, but there is room to humanize our writing and create a better experience. So, instead of leaving that to the subjectivity of each individual person on the team, the tone framework helps people get to the right tone for the right situation, for the right thing people are going through, with a much higher consistency."

Toning It Down

But what if you work on a banking app, an insurance product, or some other kind of transactional system? In most cases, you don't want to load up your product with a lot of similar sayings like, "Hi, user, how are you today?" and "That's great! Good job!" A lot of that stems from well-meaning people who really want the interface to feel friendlier. They think that by putting a couple of exclamation marks and by using the user's first name, that's going to get the user to engage with the product more.

But that's not what tone's about. Tone is present in your writing whether or not you know it. When you write in a short, neutral, terse style, that's a tone decision. Choosing not to emphasize the emotional aspects of an interface is also a tone choice.

For 70 or 80 percent of your writing in most digital interfaces out there, that's fine. Your goal is to get a user through an interaction or workflow to accomplish their goal. And because you don't want well-meaning but uninformed stakeholders inserting their own tone, wouldn't it be great to have a strategy? Even if your strategy is for the personality of the interface to be focused on helping users accomplish their tasks, make that clear to everyone involved.

DON'T ASSUME EMOTIONAL STATE

I've worked with a variety of organizations, including the following:

- A machine manufacturer
- An insurance company
- A hospital specializing in rehabilitation
- A power tool manufacturer

With each product I worked on, I had to be very aware of the situations their users could be in. After all, no one makes an insurance claim for fun.

The lesson I learned is that when it comes to tone, unless you have a lot of knowledge about the user's situation, keep it simple. People could have

Building the Facebook Tone Framework

Probst wanted to make sure that the Facebook Content Strategy team was seeing the larger picture across the wide landscape of features and products, and find patterns among them.

"So, naturally, we created a huge audit," Probst said. "This allowed us to build more tone profiles and fill in the spectrum." (See Figure 7.4 for an example of this in action.)

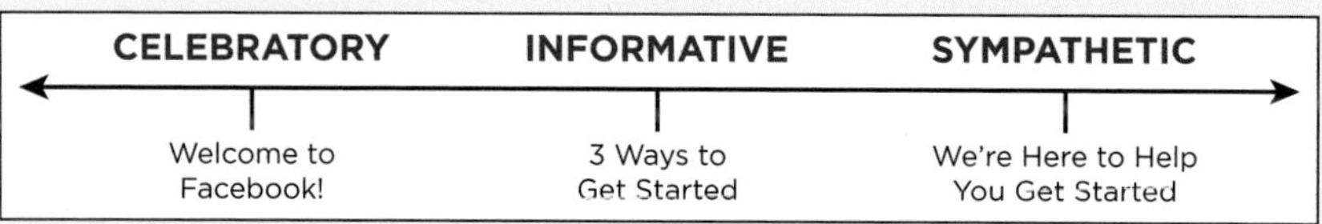

FIGURE 7.4

Facebook tones laid out on a spectrum. Developed by the Facebook Content Strategy team

They developed a series of questions to ask in different contexts, in order to develop their tone profiles. There were three main factors to consider: the interface, the user, and the scenario, or situation leading up to a message:

You (the interface):

- **Goal:** What do you want the user to do?
- **Mood:** What feeling do you want to convey to the user?
- **Archetype:** If this experience were a person, what would your mindset be? How would you relate to the user?

been going through anything when they used these products, including the loss of a loved one or the injury of a coworker.

I fielded a lot of requests to be more "engaging" or "friendly," but I always helped my teams see how that should never mean assuming too much about what users were going through. ■

Developing Tone Profiles

Tone profiles give you and your team a framework that you can use to evaluate tone decisions. There's a lot of holistic, user-centered research and thinking to do, but this is a great time to exercise your creative writing skills as well.

Them (the user):

- **Emotion:** What might they be feeling when they encounter this message?
- **Receptiveness:** How open to the message might they be?
- **Stress:** What are some extreme examples, no matter how rare, of how a user might interpret a message badly or under stress?

Scenario (the context, or situation, you're in)

- **UI Type:** What sort of UI element is this (a confirmation dialog, an error message, a subheading, etc.)?
- **Location:** Where in their user journey might this user be?
- **Next Step:** What will the user encounter immediately following this message?

"Once we saw the shape of these tone profiles, it gave us something to start testing," Probst said.

Next, the team wrote sample messages in each tone. This activity helped them evaluate how they worked in context, along with what users might be thinking or feeling while reading the message.

"That's a great exercise to do, because it lets us develop specific guidance for each tone, like 'lead with the value of performing this action,' or 'be respectful of the user's time,'" Probst said.

Start with an Audit

It's important to understand the full range of relationships a user of your product might have with you, so start with an audit of your company. Are you a large, multifaceted social network? Or are you a niche tool that does one thing, and one thing well? That will give you a good place to start, but it's invariably much more complicated than it can seem.

Auditing is just taking stock of what you have and then evaluating it. You could audit the entire experience or just a representative sample. The important thing is to have a goal: Knowing what you want to learn will help you audit with purpose.

Pick a few key workflows that your user is likely to take. If you're writing for a food delivery app, it might involve:

- Signing up for an account
- Browsing a menu and placing an order
- Cancelling or editing an order
- Getting help with a wrong order

But that's not all! In this app, there's also an interface for the food delivery people and for the restaurant that's fulfilling the order. Be sure that you're taking all of your users into account.

Take Inventory of Your Messaging

As you audit, take inventory of what you have and capture examples. It can be hard to tell what needs to be inventoried. Is it *every* word in the interface, including every button, label, and piece of microcopy? Or is it just the more messaging-heavy text, like error messages, confirmation dialogs, and onboarding experiences?

It's all of the above, we'd argue. Sure, the more word-heavy messages have the room to be more explicit in their tone, but even one- or two-word labels are making tone choices. For example, if a food delivery app calls each restaurant a "Restaurant" or a "Place," like in Figure 7.5, they're making different tone decisions.

NOTE CONTENT AUDITS VS. INVENTORIES

> What's the difference between a *content inventory* and a *content audit*? Lisa Maria Martin, in her 2019 book, *Everyday Information Architecture*, says that these terms are frequently confused: "An audit is a process; an inventory is a product. An audit is the action of reviewing a website; an inventory is the artifact that results from the audit. Please keep these terms separate! Making a distinction between process and product is incredibly useful—it helps clients and colleagues understand the difference between the grind and the output. Plus, it's easier for everyone when we're working from the same shared vocabulary."

When you're walking through those workflows, make a note whenever you see a message. It could be a simple little confirmation notification ("Congratulations! Your meal is on its way.") or as complicated as asking the user to provide detailed delivery instructions ("Tell us where your driver can find you."). It could be buried in the terminology decisions the app makes (Do you refer to the list of food as a "menu" or an "offering" or something else?), or in the way it categorizes each listing.

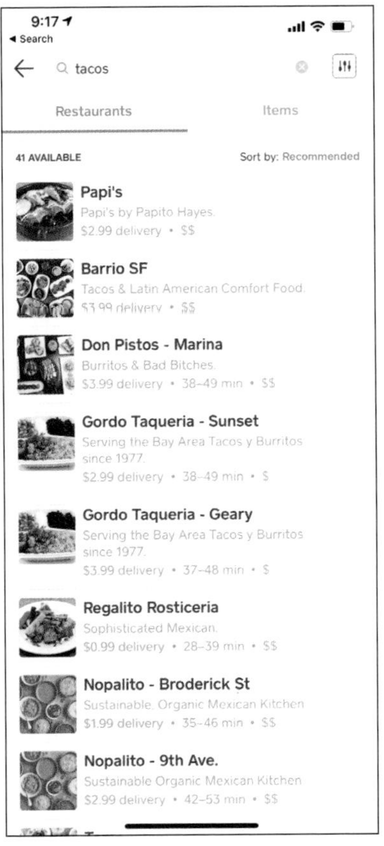

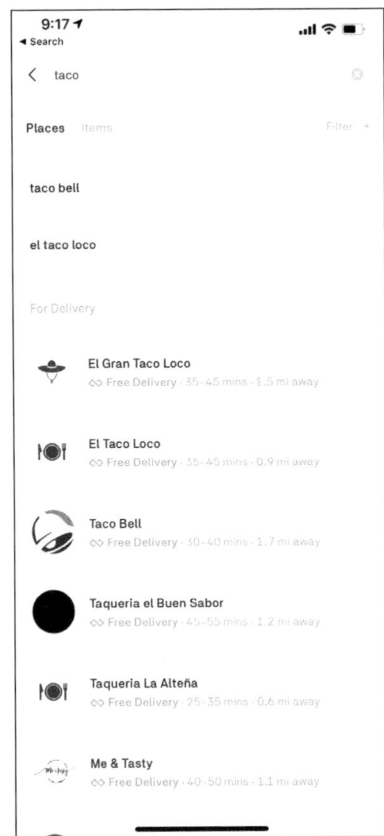

FIGURE 7.5

Side-by-side of Caviar and Postmates, one referencing "Restaurants" and one "Places."

A great way to do this: If you're lucky enough to have a large conference room (but a relatively empty wall in your house or cubicle works well, too), print those screenshots or write those messages on Post-it Notes and stick them to the wall, along with some "stage directions"—what actions your user takes to get to that message.

If you're working with a remote team, try white-boarding apps or a collaborative text document as a place where your team can capture examples and notes about them.

Show Your Work

If you work on a big team, or with other stakeholders in a different discipline, it's not a bad idea to take the results of your audit and make it compelling for others, as you can see in Figure 7.6. Audits are the *perfect* inflection point for real change in an organization, and often you'll identify other things that need to be rethought as well.

FIGURE 7.6
Content design manager Jonathon Colman created this poster in 2013 to show the process and impact of a content audit he conducted for REI, an outdoor gear cooperative.

It's surprising how often no one audits an existing experience before jumping into a new one, which shows that you've done the work and that your observations and future recommendations are holistic, rational, and based in reality.

It's especially handy if you need to justify to a manager or a product owner that you need to spend additional time on a tone strategy.

Flag the Questionable Content

Pretty quickly, you'll see some messages that aren't appropriate or don't work for the situation. Sometimes they may be poorly written or way out of context, like in Figure 7.7.

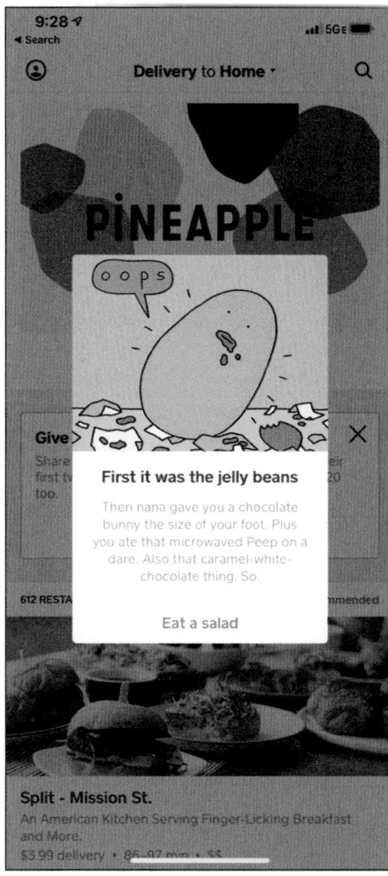

FIGURE 7.7
A modal window from meal-delivery app Caviar, shown the day after Easter. It's funny to read when you understand the context and if you're in a receptive state of mind, but generally, it's shaming the user for their assumed eating habits, and enigmatic for those who don't understand the cultural aspect of a sentence that is just "So."

Here are a few questions you can ask yourself. If the answer is "no," you may want to flag it for follow-up or rewriting:

- **Is this message contextual?** Is this the appropriate place to take the tone you're after? Was it written in a vacuum with no consideration given to where the user is in their workflow?
- **Is this message empathetic?** Sometimes, we unwittingly sacrifice empathy for writing we think is clever or catchy. We'll add extra exclamation points or make it sound quirky or delightful in order to appease our product owners or ourselves, but it might end up offending those in situations of stress. See Figure 7.7 for a great example of a message that's funny, to some, and completely lacking in empathy for others.
- **Is this message inclusive?** Does this message alienate or exclude specific populations of people? How could this message go wrong? How could it exclude people of different genders, sexualities, ethnicities, race, ages, or abilities?
- **Is this message ready for translation?** Is this message idiomatic or overly colloquial? How might someone interpret it who doesn't speak or read the language it's written in? How might the message be lost if someone translated it literally?

Group Similar Messages

Based on this audit, it's time to start thinking about tone systematically. Remember, this is how you think about the conversational quality of your interactions. You're going to try to make your writing map to real-life interactions as much as you can.

So, similar to the questions the Facebook Content Strategy team asked to develop their tone profiles, take a look at the three main factors to consider in a conversation: you, the other person, and the situation you're both in.

Take those workflows you audited, and for each one, try to answer the following questions

- **Scenario:** In what situation would a user be encountering this message?
- **User's State of Mind:** How might the user be feeling upon receiving this message? How receptive might they be?

- **Product Intent:** What's your goal? What do you want the user to accomplish?
- **Tone Attributes:** What further characteristics might this tone share with a character archetype? (Is it supportive? Nurturing? Trusted? Patient?)

Try to answer these questions succinctly (in fact, you can use a table like Figure 7.8) and look for similarities. Chances are you'll see natural similarities emerge—perhaps in the goal a user has, or in how receptive or willing they are to see a message. This kind of activity is great for answering broad questions and gaining context around the right tone to develop for individual messaging opportunities.

Scenario	*User's State of Mind*
Product Intent	*Tone Attributes*
EXAMPLE MESSAGE	

FIGURE 7.8
Here are some tone profile worksheets the authors use in their UX Writing Fundamentals workshop.

IN PRACTICE

Write in a Variety of Tones

One of the best ways to explore the ins and outs of your various tone profiles is to take one message and try it out in various tone styles. It's a great way to see the outer boundaries of the tone and if it makes sense for a message. If it looks ridiculous when you read it back, it's probably not the right tone.

Let's take an example message—a forced password reset, for example. Here's the message:

- UI Type: Confirmation dialog.
- Your password needs to be reset every 90 days.
- You didn't reset it in time, so it's now expired.
- Choose a new password and log in again.
- It must be more than 12 characters and include letters and numbers.

And here are the sample tones:

- **Encouraging:** Positively motivating you to accomplish your goal so that you can reap a value.
- **Informational:** Neutral and no-nonsense. Cutting to the heart of a message to present the facts.
- **Trustworthy:** Security and safety are our top priority, and you can trust us to keep your information safe.
- **Sympathetic:** We're very sorry this has happened to you, and we want you to know that we're here to support you in this difficult time.

Now try them out!

Look for Patterns

The purpose of the audit is to see the big picture of a user's experience through a product or workflow. It gets you out of the singular message mindset, just as it gets a more visual designer out of the screen-by-screen view of their prototyping app. As you look at each of your messaging groupings laid out on a big audit, some patterns emerge. You might notice:

- Perhaps users start each interaction feeling confused or overwhelmed by the message and the clutter on the screen, or they feel they're too busy to read all these messages.

Encouraging

Hi, Andy! Let's get your password reset—it's been 90 days since you last updated it. Go ahead and pick a new one that's 12 characters or more, and includes letters and numbers. Then use it to log in again.

CTA: Reset Password

Informational

Your password is 90 days old and has expired. Create a new one and sign back in.

CTA: Reset Password

Trustworthy

Your security is important to us, so please create a new password. Make sure that it's 12 characters or more and includes both letters and numbers. Then you can use it to log in securely.

CTA: Reset Password

Sympathetic

We're sorry, but your password expired, and you need to reset it before logging in again. We want to make sure that your account is safe, so please use 12 characters or more and include both letters and numbers.

CTA: Reset Password

After reading these options aloud, "Sympathetic" is clearly not the right fit. "Encouraging" also seems a bit overboard. "Informational" or "Trustworthy" makes a lot of sense because this is an inconvenience, but also a familiar action.

- Perhaps the intensity of a message's expression is just too high. What the user really needs is a clear, concise message in certain parts.
- Perhaps the user just finished a pretty big workflow and upon successful completion, they're satisfied that they accomplished their goal.
- Or, perhaps we made an assumption about a user's state of mind or situation, and they interpreted our message badly.

Pull It All Together

By now you've hopefully seen some patterns and noticed how the interface and the user's relationships interplay and affect each other. This is a great time to think about if the writing and the design supports or hinders your goals and the user's ability to accomplish those goals.

Taking what you've learned about what the writing does well, and how it could be better, start to put together tone profiles. This profile represents your first attempt to talk to users accurately and strategically based on context. Just as you talk differently to a parent than to your best friends from college, this is a framework that lets your product talk to new users, happy users, or stressed-out users.

Map It Out

There are so many ways to visualize your tone spectrum. Adobe's is mapped out from proactive to reactive. The more toward the middle the tone is, the more neutral and less expressive it is, whereas the tones on the outer limits are more expressive.

This visual mapping is useful because it tends to follow a typical linear path through a user journey:

- **Motivational:** Typical for onboarding experiences
- **Useful:** Great for lower-intensity expressions, like around communicating new features in an already existing experience
- **Reassuring:** Good for privacy-related and financial transactions
- **Supportive:** Good for users who are frustrated or experiencing a broken experience.

But it also shows that the majority of your tone is informational and really optimizes toward a lower intensity expression. It's the neutral, concise format that you strive so hard to achieve.

Start Using Tone Profiles

Now you have tone profiles, but they don't have value unless you have some way to tie them to the actual writing. Keeping these profiles in mind, start to write some tactical rules that follow that guidance. For example:

A motivational tone:

- When appropriate, greet the user.
- Emphasize the value of the thing you're motivating them to do or learn.
- Acknowledge that they might be potentially confused or overwhelmed.

An instructional tone:

- Recede into the background as an active presence.
- Stick to just the information you're trying to relay and nothing more.
- Emphasize what's happening at this moment in time for the user, or what's immediately following when making an action.

A supportive tone:

- Recognize that the user is under stress.
- Put your trustworthiness and the user's safety and security first.
- Emphasize the outcome, or resolution, of their workflow: What will be better once they complete their action?

Revisit It Periodically

Don't rest on your laurels. You evolve as a writer. Your software evolves and changes. If you're on a writing team, your dynamic evolves. It's important that you revisit your tone framework regularly to see if it still makes sense.

Are there big gaping holes in the middle that you're not accounting for? Or do you have a profile that's just not relevant to your user base?

You should check back in with this framework quarterly and ask yourself or your team some questions:

- Is this framework still meeting my needs? It it still helping me or my team write for the user?
- Are there new features that might present a new context to the user (for example: a new account management panel where users may be forced to reset a password)?
- Are there whole areas of your tone framework that you're just not using?

Finding What's Right for You

At the time of this writing, some tone systems frame tones in the context of positive and negative, as seen in Shopify's Polaris design system in Figure 7.9.[2] Also, 18F's tone framework is almost a mini-voice guide, with style attributes and writing principles for each, as seen in Figure 7.10.[3]

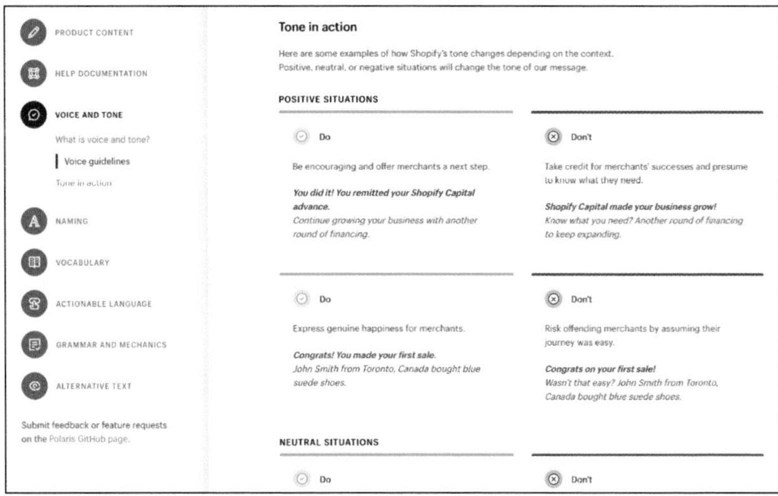

FIGURE 7.9
Screenshot of Polaris design system from Shopify's Voice and Tone guidelines.

All this to say that as with many frameworks, the way you audit, track, and plan tones may vary depending on your approach, your business, your users, and your capacity for a strategy. The more complete and planned-out your framework is, the more rigid it is and inflexible to stress cases or outliers. The more simple and broad it is, the more ambiguous it is, and the more it allows for interpretation within it.

The most important thing is that you're being strategic in the way that you're communicating with users, and thoughtfully considering their context as they interact with your interface.

2 "Voice and Tone," Shopify Polaris, October 2019, https://polaris.shopify.com/content/voice-and-tone#navigation

3 "Voice and Tone," *18F Content Guide*, October 2019, https://content-guide.18f.gov/voice-and-tone/#choosing-a-tone

Choosing a tone

As we mentioned earlier, your voice is a constant, but your tone is a variable. Consider the following: If you're having an irredeemably terrible day, you might get peeved at a store associate who chirpily (and repeatedly) asks if they can help you with anything. Instead of picking up on your nonverbal — or perhaps verbal — cues, this associate is **tone-deaf**. The associate maintained a consistently helpful voice, but they failed to shift their tone from energetic to restrained. As a result, their message (however valuable or well-intended) is lost on you.

To avoid going the way of the associate, think about your users' needs in different situations. Use these needs to determine your tone.

Let's consider three examples that target three different reader groups. Obituaries, technical blog posts, and marketing emails targeted at newly engaged couples have vastly different tones. Why? The three types of writing correspond to audiences in three highly different emotional states.

Type of writing	Intended readership	Tone	Example
Obituary of a prominent community member	People who knew (or knew of) the deceased	Respectful, reverent, somber	"Professor Pelham was respected by his colleagues and revered by students, many of whom would wake before dawn on registration day to ensure gaining entry to his classes. His wit, gentle humor, and compassion left their mark on everyone he talked to."
Blog post announcing	Developers and other	Direct, impartial	"The Open Source Style Guide is a comprehensive handbook for

FIGURE 7.10
Screenshot of 18F's tone framework.

CHAPTER 8

Collaboration and Consistency

Building Your Practice

Writing to design experiences isn't about memorizing a list of the right and wrong ways to craft a sentence. Your users are unique, and your organization is unique. Your writing should embrace that.

The same is true for your career. There's no "correct" job title or description for this work. There's not a single way you should contribute to your team or help others understand what you do. Just like you have to find appropriate words for your users, you'll have to find an appropriate way to work within your organization. A big part of your job is to figure out what your job is.

Maybe you're writing full-time—hired specifically for your writing skills to bring clarity and usability to the product (or products). Maybe you spend most of your time doing something else, like visual design, product management, or development—but want to make sure the words that make up the experience you're creating meet the needs of its users.

Whatever the case, designing your role is just as important as any of the writing you do. Remember, everything you've learned in this book so far is worthless if it doesn't affect the final product. Good writing—like good design—must be implemented and used to be of any value to your users.

Working with Your Team

Whenever you start working on a product, you may feel like everything is urgent—that you need to apply every lesson you've learned to what you're working on.

You don't. It all comes down to meeting needs. Sometimes, people don't need that cool UX deliverable you've been dreaming of creating. Don't confuse making a deliverable with making a difference.

Start by meeting the needs of your team (even if you have different priorities), and then you'll have the support you need to meet the needs of your users. There's so much that has to be written, and the pressure on product teams to deliver is constant, so you'll find that they have all kinds of things for you to do, ranging from marketing messaging to interface copy to error strings.

Make no mistake, this stuff was being written before you showed up, but it was out of necessity. Most teams have writers on them, but they aren't people with formal training. Writing happens as needed, with developers writing for the scenarios they're coding, designers writing for their interface designs, and product owners editing or rewriting a lot of it. You can lighten their load, which will win you some favor and goodwill.

This may be difficult for you, because if you truly want to design the experience, you'll need to do more than just write the bits of UI copy you've been assigned. You'll want to spend time learning about your users and solving their problems. However, pitching in is truly the fastest way to get there.

Michaela Hackner has a lot of experience joining teams and helping them see how she can help. She's a UX Content Strategist who has spent a large portion of her career making financial services accessible so that people can make better decisions about their money.

Hackner has found success by developing empathy with the team she's joining. "For most folks, they're used to handing off a document or a spec or some sort of wireframe and asking people to add content to it because that's how they think about it," she said.

After joining a team at a large bank, she was initially frustrated because the team seemed uninterested in hearing how she could help solve design problems. They just wanted her to write.

"I got in there and started doing that work because that's what they needed in that moment—and until those needs were met, it was going to be hard to find common ground," she said.

However, her contributions to the team didn't go unnoticed. In fact, she said this work created the space for her to drive the biggest business impact during her time at the organization, not to mention establishing a great working relationship with her product partners. "I found that once I had built that trust and did the things that they needed help with, I could start showing them different ways of working."

Empathy is all about understanding the emotions and motivations another person is going through. Indi Young writes about this in her book *Practical Empathy*. She points out that a common mistake in the business world is to try to make empathetic decisions without first developing empathy for the people affected by those decisions.

The same is true of joining your team. Before you can expect to revolutionize the writing, you need to listen to your team and understand the pressures they face, their motivations, and their needs. Do this by listening and asking questions, and then act on what you learned.

THE EFFECTIVE INEFFICIENCY OF OFFICE HOURS

When I joined Adobe in January 2017, I was coming in as the first-ever UX content strategist—the lone writer embedded on the 300-person product design team. It was vital that I talk to designers and product managers from across the breadth of products at Adobe.

I started to hold "office hours," a few hours set aside each week, where design teams could book 30-minute slots with me, and I could try to help them solve design problems using words.

I wasn't prepared for how popular my office hours would be. It turns out, many of our product teams knew they needed better words in their interfaces. And I quickly realized that half-an-hour was not nearly enough time to give them much more than a hot take and a couple written-on-the-fly text strings.

But it *did* help in a few other ways. It allowed me to see the big picture and identify widespread problems across our product landscape, where I could make a difference at scale. That was invaluable in helping me prioritize my work and stem the duplication of effort that many teams were facing within their organizational silos.

It also let me meet many, many designers, researchers, and product managers, which let me develop relationships with them and evangelize my practice more effectively, which eventually led to getting buy-in for headcount and building my team.

If you're one or among a few writers on a much larger team of designers, consider office hours. More often than not, you won't immediately solve the problems they'll bring to you, but you'll be empowered to solve much larger, more systemic ones. ■

Design Your Process

A lot of the frustration that people experience when writing for a product team is that things will come to them too late.

What things? Decisions, requests, and generally most of the interesting, strategic work. It can be frustrating, but remember: Writing looks easy to the people who aren't doing it.

If the process you're working in makes writing difficult, change it. Rather than staying frustrated, you can design a workflow that enables you to do better work.

Here's what this may mean:

- **Planning in advance:** Last minute, reactionary work will generally be your worst. Do what you can to understand the product backlog as a whole and take a plan to your team that gives you space for the writing.
- **Calling out blockers:** In the development world, a *blocker* means that you don't have what you need to get the work done. Need more information? Waiting on someone to get back to you with a decision? Tell your team.
- **Acknowledging all the work:** Need to review your writing with a stakeholder? Does legal need to sign off on what you create? Do you need to schedule wider meetings about the work? Do you need to do research? All these things are part of doing your job, and you should factor them into your timeline.

Give yourself the space to design a better workflow. Spending a few hours on it before you get very far may save hundreds of hours in the long run, and you'll do better work as a result.

Whether writing is a full-time job for you, or you're fitting it into other responsibilities, you still need time to do it. When writing is designing, it involves a lot more than just writing, and we're not the only ones saying this. Here's what Scott Kubie said in his book *Writing for Designers*:[1]

> ...writing is more like design than you might think. Common design activities like framing the problem, identifying constraints, and exploring solutions are part of writing, too. Many of the methodologies one might use in UX work can be part of a writing workflow: stakeholder interviews, user research, content auditing, ideation workshops, critiques, and more.

Designing your role and explaining it to others will help change that mindset, setting you and other writers up for success.

1 Scott Kubie, *Writing for Designers* (USA: A Book Apart, 2018), introduction, Kindle, https://abookapart.com/products/writing-for-designers

Invite Yourself

Being left out of important conversations can be difficult and discouraging. Maybe it's a kick-off meeting, a sprint planning, a feature ideation session—whatever. You're left feeling like you should have been there because the outcome has a big impact on the product and on the users you're writing for.

More than likely, the people leaving you out of the conversations simply don't realize that a writer can make a huge impact early on in a project.

So what should you do? Start inviting yourself.

This is challenging and a bit scary for anyone, and it's especially difficult for introverts. But as hard as it is, it's going to be much harder for you to go through the emotional journey of wanting to be invited, and then finding out later you weren't invited over and over again.

Here are a few tactics for getting an invite to a key meeting:

The Listen-In

> "Hey—it sounds like you're making some important decisions in that meeting. Can you add me just so I can listen in?"

Use this dialog when you're dealing with big egos or people for whom hierarchy is important. They may feel they've earned a spot at that meeting and could be hesitant to let others in.

You can mix it up by saying things like "I'd just like to be a fly on the wall," or "I just want to make sure I stay informed about what may affect my work."

Most of the time, this approach works wonderfully.

Sometimes, it doesn't. A product owner or manager might say, "Nothing personal—we're trying to keep the room small so we can make quick decisions."

That can feel devastating. Remember, this isn't personal. You could respond with, "There's a lot of high-impact language in this interface, and if I'm there to guide decisions around language now, we can save so much money and time down the road once these designs start getting fleshed out."

Which leads right into the next tactic:

The Productivity Enabler

> "It seems like that topic relates to what I'm working on now. Could you add me to the meeting? I'd hate for any changes to come late and block the dev team."

If you ever need to get people to add you to a meeting without question, mention that it could affect the progress of the dev team.

It's silly, but lots of companies act like there's nothing worse than a development team with nothing to do. By pointing out that changes could affect you and them, you'll help people see why you should be involved in early conversations.

The Helping Hand

> "I know a workshop activity we could do to get the team aligned. Are you open to me facilitating that for the group?"

Whether it's strategy statements, affinity clustering, or a design studio, people love workshops. They're engaging, they focus the conversation, and teams often lead with something tangible.

When you're comfortable leading workshops, an incredible way to get involved in strategic conversations is by offering to facilitate them and lead the group to a shared understanding.

Create Contagious Transparency

Sometimes, it's hard to get involved because other stakeholders just don't know about the work you do.

You may see a button label and think of five questions about what happened before the user sees it. Meanwhile, the developer who asked you about it sees the same button and thinks that if they can get you to write the label for it in the next five minutes, they'll be able to push their code to the test server before lunch.

Rather than wrestling with those five questions in your head, get them out in front of everyone. Post them on the user story in your company's Jira instance (a commonly used tool for tracking issues and projects). Mark the story as blocked until you get answers. Ask about it on Slack or Microsoft Teams, workplace chat apps your organization might use.

Transparency helps everyone understand your workflow and how they can help you.

But it's not just about showing people what you're going through—it's about helping others understand why your questions are important to the product.

By being transparent, you'll start to give others the confidence to do the same.

Ask Questions

If you care about words, you'll likely find yourself asking a lot of questions about what you're working on. This is because good writing makes things clear, and often, if no one has been focused on writing, things aren't very clear.

The people you work with will look at you strangely, as if you're a species of animal they had never encountered before. They may even tell you not to worry, and that you really shouldn't ask so many questions.

But questions are important.

We ask questions to understand the world around us. For many teams, no one has tried to understand the words that make up their product before. It's threatening and unfamiliar to them.

We can guarantee that many questions have been asked about code deployment, product-to-market fit, roadmaps, CSS rules, database queries, APIs, and a host of other details that make up your team's activity.

Asking questions about the product and its purpose greatly increases your effectiveness. Go beyond asking about what happens in the interface and start asking about the problems the product is meant to solve. If *you* don't understand what's going on, how do you think your users will feel?

Hillary Accarizzi's experience has led her to build a culture of empowerment on her team, where writers are comfortable asking questions. She manages writers and content strategists who work on digital products for a large insurance company.

Accarizzi tries to help each writer she manages get comfortable contributing to the overall experience of the product. "I push them a lot not to just start on what's given to them," she said. "I want them to understand that their value is not limited to the words."

One of the big ways she does this is by encouraging team members to ask questions. "If someone isn't comfortable doing that in big meetings, find someone to trust and ask them privately," she said. "The more questions they ask, the more comfortable they'll feel doing it; the better the work will be; and the more they'll be seen as an expert."

Accarizzi notes that it can be challenging to contribute to a product as a writer because of how much advocacy for your own work is involved. "You do have to be vocal, you do have to speak up, and you do have to show the value," she said.

However, it has been rewarding for her. "Our entire lives are digital," she said. "If we can help somebody have an experience that is so smooth and so rewarding and fruitful for them that they don't even think about it, that's the goal."

Often, you may be presented with an interface design by a designer or a product owner, and asked to "please wordsmith this." There are likely so many questions swirling around your head, and please, ask those questions!

Here are some questions you might have about a specific interaction you're working on:

- What's the goal of this workflow?
- What is the user trying to accomplish here? How does this design help them?
- What did the user see right before this message?
- What does clicking the "OK" button do?
- What are the steps a user took to get to this point?
- Do our users understand what this means?

And here are some questions about the product:

- What's the intended outcome?
- How does our business benefit?
- How are we measuring success?
- What's our vision for this product?
- What behaviors are we trying to encourage?

There might not always be clear-cut answers to these questions, but asking them benefits the whole team and gives you clarity.

Collaborating on Writing

Every organization is different and has different needs. One may hire dozens of writers or content strategists to focus on their work full-time. Others may spend years with one person leading those efforts.

However, even if you're the only person focused on writing, you can still form a community of support for anyone who writes the words that your users interact with. Most organizations include people who care about writing and want to do it well. Your job will be easier if you build a connection with those people. It's invaluable to have someone you can bounce ideas off of, or help you reach out to stakeholders outside of your usual department.

On a practical level, anyone who wants to do this type of work benefits from a mindset that propels them forward and energizes them. You can create that energy by working alongside other people who share your goals.

It helps to establish shared spaces where people can bring their questions, thoughts, and ideas. These could be physical, if you can swing it, like a room or a wall dedicated to showcasing recent work and resources. It could also be a digital space, like a room or channel on your organization's group messaging tool or an email distribution list, like this cross-organizational Slack channel in Figure 8.1.

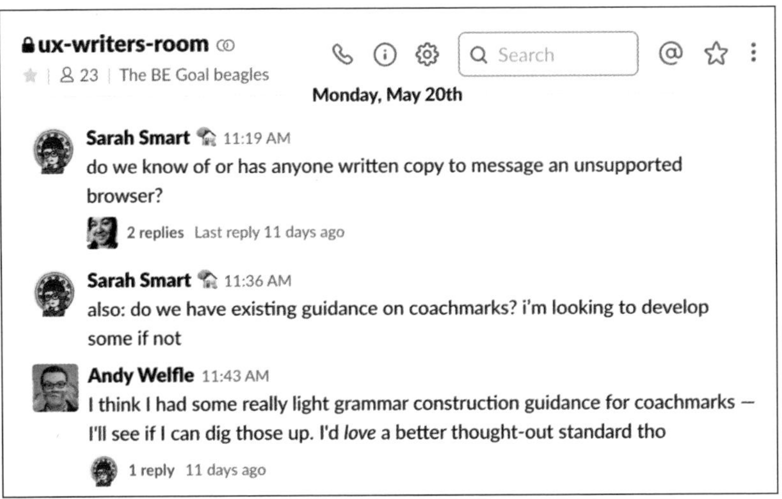

FIGURE 8.1
Adobe Design's #ux-writers-room Slack channel.

Apply your design skills to the creation of these spaces, so that they're easy to be a part of and beneficial to everyone involved.

Making an effort to connect like this will help you and others see that you're not in this alone—you've got plenty of people around you who are trying to solve similar problems—and you can help each other solve those problems.

But this type of connection is about more than not feeling alone. There are practical ways you can help each other do great work.

Facilitate Understanding

At the end of the day, regardless of your title, there's one thing that's important to remember: Everyone on the team is working together to design the experience.

This message is especially hard for people with "designer" in their job title to hear, but it's true.

Jared Spool, an expert in the field of user experience, describes design as the rendering of intent, and points out that everyone on the team is a designer:[2]

> Having everyone on the team act as a designer is not, by itself, a bad thing. It brings a lot of power, knowledge, and experience to bear on what are often complex problems to solve. In fact, it can be a great thing when the team has the same intentions.

Since writing is designing, facilitation is a critical skill. You can help your team figure out what their intentions are. This is a critical part of what it means to design something.

Facilitation is the act of bringing people along with you as you explore an idea, explain a concept, propose design changes, and more. Without facilitation, things just happen to you. With facilitation, you have the opportunity to apply design to every moment of every day.

A lack of facilitation will leave you feeling ineffective, and it will leave your colleagues feeling like you're out of touch. A good facilitator helps other people feel involved, excited, respected, and productive.

---

2 Jared M. Spool, "Design Is the Rendering of Intent," UIE, December 30, 2013, https://articles.uie.com/design_rendering_intent/

Let's say you're getting ready to redesign the login experience for your product. You're excited, because you know a lot about this topic—including where your users are struggling and how it can be improved. You prepare your data, design a solution, then excitedly present everything to the team only to find that they disagree with your assessment of the problem and don't like your solution.

This kind of experience can be very frustrating, but you can often prevent it through facilitation.

Instead of jumping right into a solution, take things in steps:

1. **Identify and document the problem.** Or at least, what you think the problem is. Figure out what problem you're trying to solve and why you need to solve it. Gather evidence: screenshots, testimonials, reports, and other artifacts, and organize them in a coherent narrative to present to your team.
2. **Agree on the problem.** Start by listening to what your team thinks the problem is and capture the conversation for everyone to see. This will help you sort out disagreements early on and give you an opportunity to reference your research without people feeling like you're making a case.
3. **Establish objectives.** Once everyone has a shared understanding, people may see more clearly that there's work to be done. Write down your team's objectives in a shared place. Maybe you can capture them as acceptance criteria in a user story, or maybe you can send out an email listing them after your meeting. Whatever the case, you want to move from having your own objectives to having shared objectives with the team.
4. **Design a solution rooted in objectives.** Now it's time to create your solution, and it will be so much easier knowing that your team has a shared understanding of the problem and shared objectives. It also gives you something to point to outside of your own opinion when explaining your solution, and the team is far more likely to adopt it.

These steps can happen quickly, and the design solution you end up creating may be the same whether you follow these steps or not. The difference is that careful facilitation makes your solution much more likely to be adopted, developed, and put in front of your users—and that's what really matters.

CANVASES AND WORKSHEETS

At my first proper UX job, I worked with Scott Kubie as a content strategy and writing duo. He introduced me to the idea of using canvases and worksheets to create shared understanding.

We realized early on that agreeing on the problem was a critical first step toward getting anything done. At first, we tried asking questions directly, but we noticed that teams would get defensive and guarded when we did this. They felt like these two people who were relatively new to the company were interrogating them about business problems and user needs.

By making a simple artifact that captured the information we needed, we created an environment where the thing (a worksheet or canvas) was asking for information and the team was filling it out together. It created an environment where everyone involved felt like a team, and we moved from being perceived as interrogators to facilitators.

Figure 8.2 shows a worksheet I use with my current team. It captures the basics like the design problem, user needs, and business goals, but it also enables teams to make notes about their discovery and design activities.

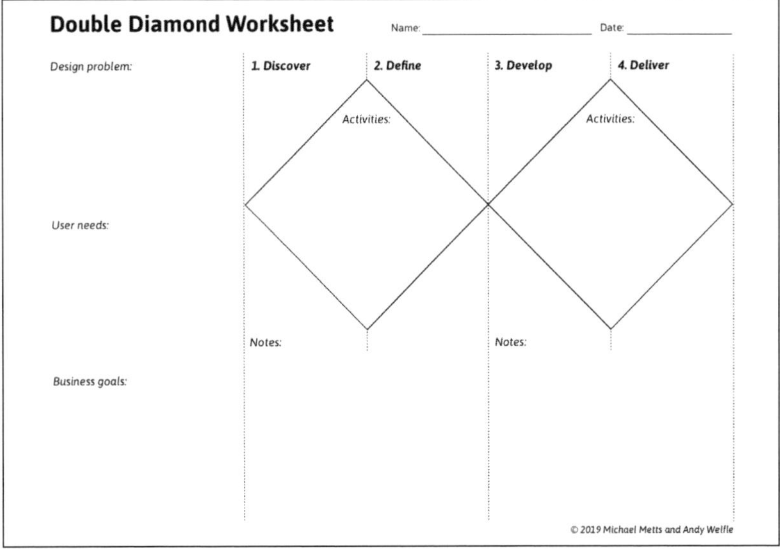

FIGURE 8.2
This Double Diamond worksheet helps a team track their discovery and design activities together.[3]

3 "What Is the Framework for Innovation? Design Council's Evolved Double Diamond," Design Council UK, October 2019, https://www.designcouncil.org.uk/news-opinion/what-framework-innovation-design-councils-evolved-double-diamond

Here's the secret: You should always have a goal. You don't have to share this goal with anyone (although anyone helping you facilitate a conversation like this should certainly know).

In this case, my goal was to give teams a warning signal if they hadn't spent enough time in discovery. The team I was working with often jumped into implementation too quickly, creating solutions without a clear user need. This worksheet served as a mechanic to highlight when more work was needed in discovery without making people feel like they were being attacked. ■

Show Your Work

One thing visual designers figure out pretty quickly is that the visual aspect of their work is powerful. As a writer, you should leverage that power.

We're not saying you should learn color theory or put together a mood board. What we're saying is that you should use visual tools to help other people understand your work. Visual artifacts are a critical part of your role because they allow other people to process ideas more quickly and retain concepts more easily.

Don't be intimidated by the idea of using a design tool. Instead, focus on the story you want to tell. Once you know what you want to show your audience, you can always learn how to visualize.

Tools are easy to learn. Clearly communicated ideas are much harder to come by.

Maps

Maps help people quickly orient themselves to a system or a thing. They're often just as useful for providing personal clarity as they are for your team. If you've ever found yourself craving an understanding of how your work fits into the larger plan, you've probably just identified the need for a map.

Figure 8.3 shows a map for a restaurant's mobile app that was created using an object-oriented UX approach. (In this case, rather than diagramming a fully realized template, you're diagramming the system of *stuff* that might be in that template.) The idea is that you map the system with an understanding that it's not a strictly linear experience. Users could interact with different elements at different times in their journey for different purposes. The cart appears on the screen only when something is in it, for example, but could appear on any screen in the app.

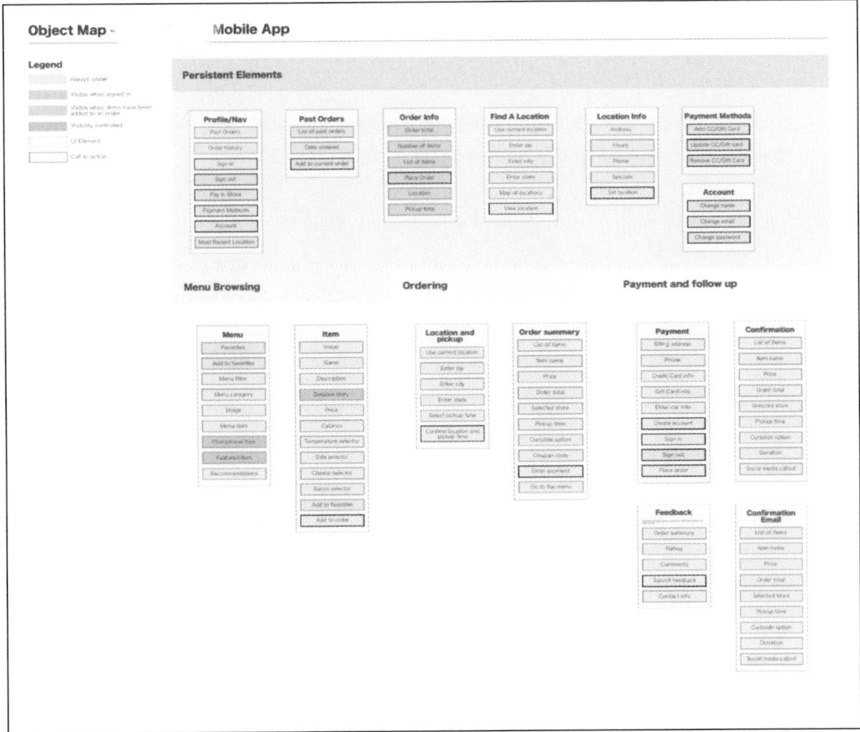

FIGURE 8.3

An object map showing all the different screens and states for a mobile app. The color coding shows when elements are visible.

Flows

If your users have multiple options that will lead to different experiences, a great way to show those diverging paths is through a flow.

Flows are especially suited to linear logic. For experiences where the user and system take turns talking to each other (like a conversational UI), flows can be an especially good way to help people see how the words in an interface change based on the user's path.

They also help for showing how a process could work, before adding the interface or language. Figure 8.4 shows a flow for how to support customers who are having trouble logging into a system.

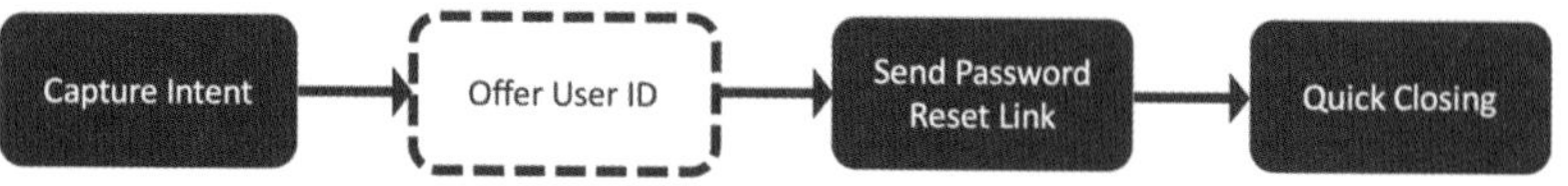

FIGURE 8.4
A flow for a conversational UI that helps users who are having trouble logging into their accounts. Courtesy of Katie Lower.

Interface Designs

Often, people reviewing your writing will want to see what it looks like in context. This makes complete sense, because for visual interfaces (pretty much anything with a screen), words and interactive elements work together to create the design.

Learn to use the latest design tools and share files with your designer. You should familiarize yourself with the design tools that your team uses—Sketch, Figma, Adobe XD, or even Photoshop are popular around the time this book was written. Often, design teams will have files for design software filled with standard elements as part of their design system.

It doesn't have to be perfect, but showing an example of your writing in context, or how it could be implemented, can be incredibly helpful for stakeholders focusing on the broader context.

Critique

Critique can be a great way to create visibility for the work you're doing, and for demonstrating how difficult and complex it can be. In a world where people practically celebrate when a meeting is cancelled, critique is one of the most meaningful and productive ways you can use an hour with your coworkers.

Critique should be energizing to you and your team. It should be an opportunity to hear a lot of different perspectives, and the work should be better afterward.

However, if you want critique to be beneficial, you need the right mindset. Here are some important things to keep in mind:

Focus on goals and users, not opinions. It doesn't matter if someone doesn't like the way you wrote something. For their critique to be valuable, they should point to how doing things differently could meet users' needs or business goals better.

The work should be in-progress. Critique should happen early enough in your process that it doesn't feel disruptive. Bring work that's messy, typos and all. It should be polished enough that people can understand it, but not so complete that changes would require serious rework.

Let the creators scope the feedback they want. Ask someone before they present, "What sort of feedback are you looking for?" It empowers them to control the conversation and avoids a feeling of vulnerability, like they've been thrown to the wolves. Perhaps they're just looking for notes on the word choices they've made. Sometimes they just want an eye on the interactions, trigger points, and entry/exit points they've worked on. And sometimes they just want a broad, general gut-check. For some critique-givers, though, it can be hard to scope or limit their feedback. The facilitator should make sure they respect the presenter's scope.

If you created the work, you can critique it, too. It's a common misconception that if people are critiquing your work, they're critiquing you. It should never feel that way. Everyone should be critiquing the work with your goals and users in mind. This means you can turn around and critique your own work along with the group.

RUNNING REMOTE CRITIQUES

One of my favorite ways to run critique is in a remote session where everyone has access to the prototype and can look through it at their own pace.

Commenting functionality is a way to enable those who may not be comfortable speaking up in front of a large group right away. They can leave a comment with their thought, and then I can ask them to expand or provide more detail if needed.

I also keep critique guidelines on the screen the whole time. Figure 8.5 shows what's on my screen during a remote critique. ■

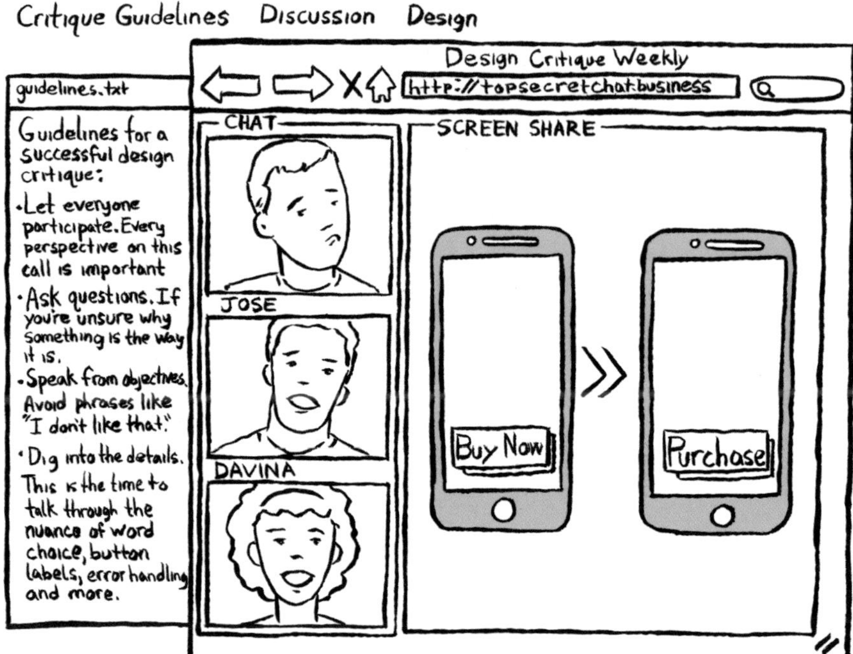

FIGURE 8.5
Critique guidelines and a discussion area are included in remote critique sessions.

Creating Consistency

As you collaborate and work with other writers, you'll begin to see that people approach things differently and write differently. This is where you'll want to start focusing on consistency.

Consistency is important because sometimes teams spend time solving the same problem over and over again. It's inefficient and expensive for the organization, because different members of different teams spend time designing and building the same thing in different ways.

It's also cognitively expensive for your users, who have to spend their time and energy learning new ways to use and recognize your products.

Identifying and replicating ideas that align your strategy and work for your users means the organization can leverage the thought and intention that a few people have already spent on a design element. This process makes everyone more efficient and the product more consistent.

In the design world, maintaining consistency seems like a foregone conclusion. It's given as a justification for almost any design decision, and leaders frequently push for it. However, consistency comes with a warning label: Without a *strategy*, any effort you make to be more efficient and consistent will just help you do the wrong things faster and more often.

REUSABLE PATTERNS

When it comes to words in particular, I try to identify patterns: ways to solve a problem that anyone can adapt or use. Great patterns aren't about providing exact wording. In fact, sometimes it's better not to focus on the wording at all and spend your time figuring out the larger design problem.

A recent example involves a team I was on that was designing a virtual assistant design to answer challenging questions for its users. During testing and research, we found that users weren't always confident that the system was giving them the right answer. In fact, the system was new and had limited capabilities, so in many cases, users were getting the wrong answer, but they didn't have a way to indicate that fact to the system. In Figure 8.6, you'll see an example of how we developed a pattern that gave them a summary of the answer right away, along with the ability to navigate away if it wasn't the topic they were looking for.

We could have documented this pattern as-is, saying that you should use the "Ready to get started?" question to figure out whether we matched the user's intent. We could have also said that you should use "Ready" on a button for the user to indicate that they wanted to move forward with the conversation, and the words "Not what I need" on a button to indicate that we got the intent wrong.

If we had done so, this pattern would have had a short lifespan. Not every interaction was as long and involved as this one, so it wouldn't have been a good fit for situations where the user was getting a short answer anyway. In those cases, we should have designed the bot to clearly say what it thought they were asking and skip the option to leave the conversation.

Those reply options work great for a walkthrough-style conversation, but they wouldn't work as well for a situation where the bot is providing reference material or presenting multiple options to the user. On top of all that, buttons wouldn't work for situations where there were many divergent paths right at the beginning. In those cases, we'd need to use a different interface element or rely on natural language to move the user forward.

We ended up documenting the overall design problem without focusing on the specific language (Figure 8.6). The thinking and research that went into the design was far more important than the particular wording or interface elements, and the pattern became much more useful when we shifted our focus. ■

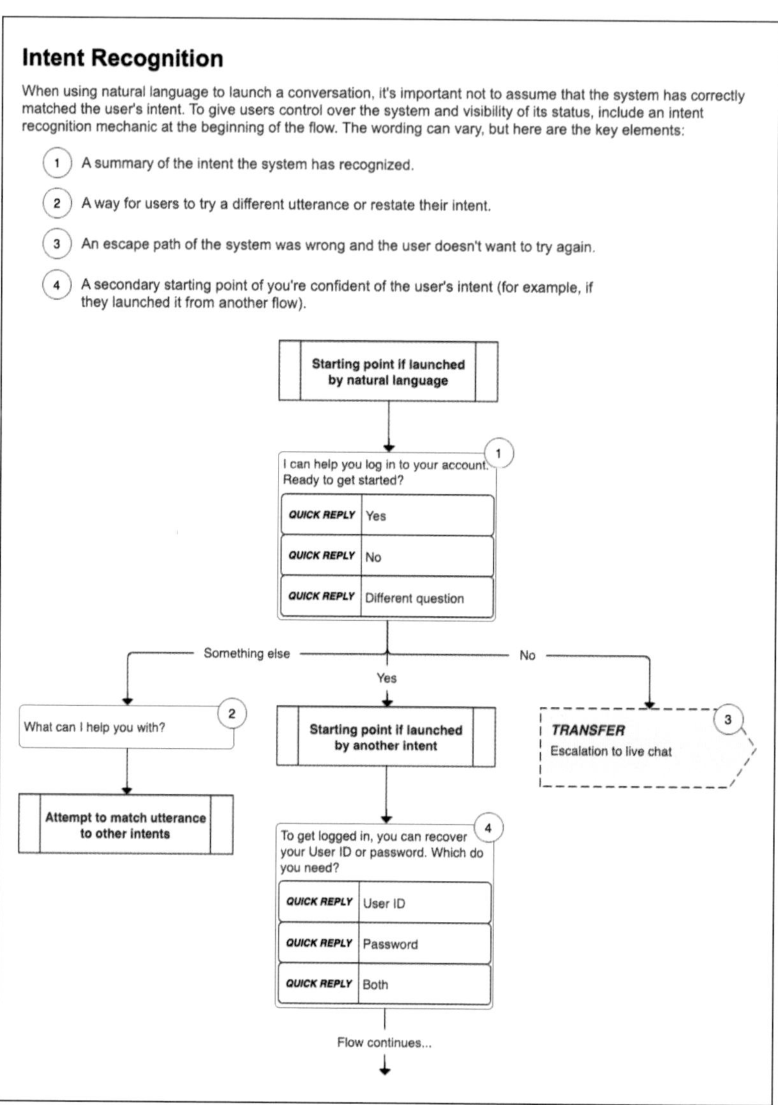

Intent Recognition
When using natural language to launch a conversation, it's important not to assume that the system has correctly matched the user's intent. To give users control over the system and visibility of its status, include an intent recognition mechanic at the beginning of the flow. The wording can vary, but here are the key elements:
1 A summary of the intent the system has recognized.
2 A way for users to try a different utterance or restate their intent.
3 An escape path of the system was wrong and the user doesn't want to try again.
4 A secondary starting point of you're confident of the user's intent (for example, if they launched it from another flow).
Starting point if launched by natural language
1
I can help you log in to your account. Ready to get started?
QUICK REPLY Yes
QUICK REPLY No
QUICK REPLY Different question
Something else
No
Yes
2
What can I help you with?
Starting point if launched by another intent
3
TRANSFER
Escalation to live chat
Attempt to match utterance to other intents
4
To get logged in, you can recover your User ID or password. Which do you need?
QUICK REPLY User ID
QUICK REPLY Password
QUICK REPLY Both
Flow continues...

FIGURE 8.6

Documentation for an intent recognition pattern used in conversational interfaces when the system has low confidence of a match.

Style Guides

One way to keep things consistent is by using a style guide.

A style guide makes sense, because in the world of print, they have served as a way to maintain consistency and quality for years. We could all just agree on which word to use for a specific situation, put it in the style guide, and then spend time on things more deserving of our attention.

In the U.S., the news industry relies on the *The Associated Press Stylebook*, lots of book publishers use the *Chicago Manual of Style*, and some academics have to suffer through MLA. The differences between these guides cause intense debate, which is a sign that people care about them and find them useful.

If your writing team isn't already using a style guide, just adopt one of the great ones out there already. For digital work, it's worth tracking down a copy of the *Yahoo! Style Guide*. Unlike a style guide that has its roots in print, it speaks directly to people working on interfaces with guidance for writing link text, error messages, and more.

Some companies will opt to make their own. If you go that route, be aware that it could be very time-consuming. Be careful committing to something like that, because you could be spending your time on more strategic work.

The important thing is to pick one and stick to it. You can answer a lot of common questions about the mechanics of writing by using one—such as capitalization, punctuation, and word choice.

However, when it comes to writing for interfaces, a traditional style guide can only take you so far. When you use words to design, even simple decisions are filled with complexity. You'll need to venture outside the world of pure writing and start to look at how words fit into the components of an interface.

Design Systems

Once your team agrees on style basics, is clear on strategy, and has identified some patterns that work for your users, you can focus on consistency and efficiency knowing it's time well spent.

A recent—and very positive—trend is to gather all the resources that a product team might need in a single place, usually called a *design system*.

The best design systems are focused on meeting the needs of anyone who contributes to, or needs to access, design. They embody the vision your team has for the experience, and include resources that help people get their jobs done, whether they are designers, engineers, product managers, or executives.

This design system could include:

1. **Strategic vision for the product or organization**, along with practical tips on how to apply it. Design principles and voice and tone guidelines fall into this category.
2. **Style guides for words and visuals.** Things like colors, typography, abbreviations, capitalization standards, and more.
3. **Design patterns made up of reusable assets and code.** Everything from files and libraries you can load into design software to snippets of code that engineers will use to build the final product.

Design systems work so well because modern interfaces are made up of lots of smaller components. Put them together and you get software interfaces. The people who create these systems tend to think of them as LEGO blocks. When you have enough of them, you can build anything you want.

If you're focused on the words, you'll need to think about how they fit into these components. If you think the visuals, interactions, and code that make up a design component have nothing to do with your work, remember that words are what give them meaning and make them functional.

It's like building something with plain blocks instead of LEGO blocks: You can build it, but your options are limited, and one bump could cause the whole thing to come crashing down.

Words give the elements of a design system meaning and utility. If your organization tries to establish reusable design patterns without considering them, you'll sacrifice clarity and usability for the sake of efficiency.

The best design systems go beyond patterns and equip teams with the context they need to make consistent, strategic design decisions. This might include documentation (made up of words), explanations about context (also words), and version histories (which should also involve words).

In Alla Kholmatova's book *Design Systems*,[4] she talks about the idea of functional patterns. These patterns determine how the interface works: form fields, drop-down menus, cards, and more fall into this category.

4 Alla Kholmatova, *Design Systems* (UK: Smashing Magazine, 2017).

Kholmatova talks about how you should focus on the purpose behind each pattern—it's a great way to make sure that each discipline practitioner is contributing in a meaningful way. Ultimately, that makes users more successful.

> The purpose determines everything else that follows: the structure of the pattern, its content, its presentation. Knowing the purpose of the pattern (knowing which behaviors it's designed to encourage or enable) can help us design and build more robust modules. It can help us know how much a pattern can be modified before it becomes something entirely different.

Kholmatova helps us see that a system like this is more than just a collection of components. You need a purpose, and that purchase should be driven by your strategy.

A Model to Follow

Shopify has done a great job integrating words into their design system, called *Polaris*. Shopify is an ecommerce service with a myriad of users, from the merchants who sell on the platform, to the people who shop there, to the developers who create integrations for it, and beyond. Shopify uses Polaris to pull together everyone at the company involved in design. And, luckily for us, Polaris is documented publicly for anyone to see.

Content strategist Selene Hinkley was one of the system's earliest contributors, and she played a major role in bringing it to life. Open up Polaris's website (Figure 8.7), and it's clear that Shopify understands the role of words as a design tool.

You'll see four things in the main navigation:

- Content
- Design
- Components
- Patterns and Guides

Having content front and center is an intentional decision by the team, as they work to build a design mindset at Shopify that includes words. "I think of information architecture as a really powerful tool for shaping human behavior," Hinkley said. "How we organize information affects how organizations operate."

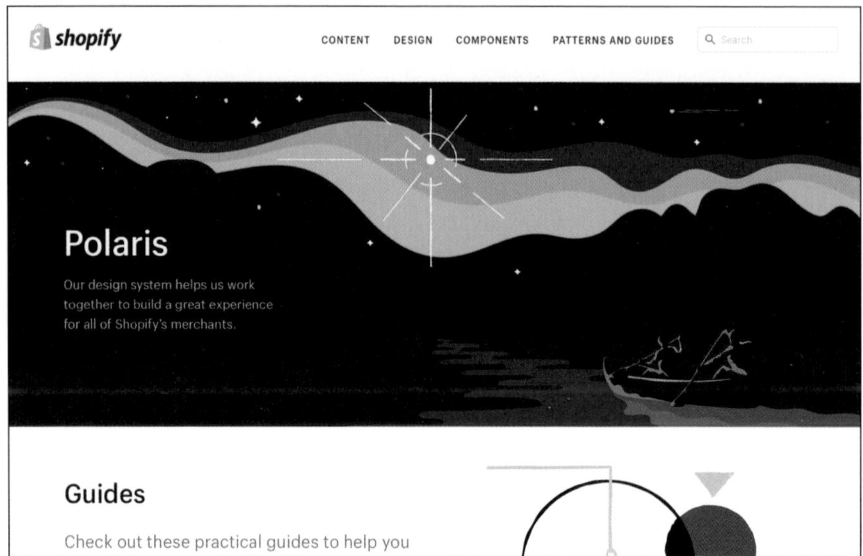

FIGURE 8.7
Shopify's Polaris design system.

One way the Shopify team made sure their design system was working for writers, designers, and developers was through usability testing. Early on, they put design and content information first on component pages, but found that developers looking for code became frustrated.

In response, they flipped the hierarchy of information, while keeping the cross-disciplinary guidelines. "In version two, the component page has the picture at the top, the code at the top, and the content and UX stuff at the bottom," Hinkley said.

They found this was more in line with the needs of their users. "It still fits with the mentality that when you need to ship fast, what you need is there," she said. "When you have more time, you can scroll down and read."

The teams building design systems like this may have ideas about how they should be designed and structured, but the Shopify team knew that a system has to work for its users if you want it to thrive.

Beyond code and components, Shopify's design system includes topics that provide strategic direction. One example involves guidelines for naming features and products. It helps teams come up with names that are on brand, proven with research and marketing insights, and consistent across the wide landscape of Shopify products.

Polaris is an impressive example, but Hinkley encourages anyone trying to get support for an effort like this to focus on value. "Don't invest too much energy at the beginning trying to convince absolutely everyone of why it's a good idea," she said. "It's exhausting, and it doesn't always get you very far."

Instead, Hinkley encourages teams to focus on where they will provide the most value. "Identify your biggest inconsistencies and prioritize those to add first," she said.

By focusing on what matters most, you'll build something your team needs, and in turn, they'll build a useful and consistent experience for your users.

Start Small

Like Shopify, some large companies have large, dedicated "Design System" websites. But if you're just beginning to put one together for your company, you don't need to start out that big. In fact, your team may never even need something like that. A shared document could do the job just as well. The important thing is that your team has what they need, and that it's easy to reference and use over time.

Your greatest enemies when trying to build a system are:

1. Feeling overwhelmed by all the things out there worth documenting
2. Feeling like no one cares about the work as much as you do

It can be tempting to try to solve every problem and think of every use-case. Fight this temptation.

Spend some time listening to what people are struggling with. Is there a common issue that comes up again and again? What do the teams you work with spend a lot of time debating and discussing? That's exactly where you should start.

To survive, your system needs to solve a problem—or multiple problems—for its users. Each person who interacts with it will need to find value; otherwise, they won't have a motivation to come back.

Amy Chick recognized this early on while leading content strategy efforts for a large telecom company's design system. She's a strategist and designer with experience in a variety of industries.

"A design system is a product," she says. "It's a tool for a specific set of users under specific circumstances. Just like how documenting business processes does not create a product, documenting components does not create a design system. You need input from the people who will use it."

Chick's first priority was to establish a shared language for the main users of their system: designers, writers, engineers, and product managers.

She facilitated a workshop with team members from each of these groups. They began by putting a collection of interface components on the wall (like tooltips, dialog boxes, and other common pieces of interface), and then going through them one by one. They talked about how each component was used, what components impacted others, and what was important to each person. Then they created a shared pool of agreed-upon language that was later used as the foundation for the naming conventions. They also grouped the components to account for platform-specific variations that surfaced during discovery.

"After the workshop, we reworked our naming and grouping conventions to reflect a system that we could all live with," she said. "That way, we could collaborate on the details of product development without talking in circles around each other."

Immediately, the team realized how beneficial this was. Not only did it bring clarity, but it also helped engineers and designers estimate the level of complexity and impact and prioritize more effectively, and gave everyone else insight into the complexity of that particular component. For example, an attending content strategist could better communicate the kinds of language considerations a component warranted, and how each fit into their overall product content strategy. It also created more awareness around the importance of content in the design system.

By focusing on a shared understanding first, Chick was able to move quickly and get lots of key people involved. "We were able to start really small so that it wasn't intimidating or risky," she said.

Finding What's Right for You

Writing the user experience depends on collaboration. It involves being nice to other people while advocating for your work. But as important as it is to be nice to others, it's even more important to be nice to yourself.

Your skill as a writer isn't defined by other people. There will always be people who don't understand your work or have a negative opinion about it.

But their opinion doesn't change the truth that without words, there would be no experience for the user. Words are at the center of how an interface is understood by its users. Your work as a writer affects the usability and usefulness of the product.

Believe in your skills and in the work you're doing. There are real people out there relying on words to navigate and use software every day. Designing your role enables you to create a better experience for them.

CONCLUSION

Throughout this book, we've emphasized that you'll need to find what's right for you. That's because you're the best person to tell your team, the management at your company, or the people at a tech conference that *writing is designing.*

We encourage you not to treat this book as a set of strict rules. Instead, think of it as a collection of ideas you can adapt. What's most important is that you stay focused on the people who use what you're making.

As you saw, the concepts in this book didn't just come from us. You've heard from many voices. Each of these people had different job titles, situations, and stories. Like them, you have your own story.

A focus on the user—on the people who experience what you create—is what sets this type of writing apart. Your words create that experience. Together with your team, you make those moments when your users interact with your product better or worse. Clear or deceitful. Useful or manipulative.

It's your choice. This book gives you the tools and methods that will help you put your users first. Now, it's up to you.

You've got this.

INDEX

A

B

C

F

G

H

I

J

K

L

M

Q

R

S

W

Y

ACKNOWLEDGMENTS

A Huge, Heartfelt Thank-you

Michael would like to thank:

My wife and best friend Karina, who generously gave her time and encouragement as I wrote. This could not have happened without her sacrifice and love.

My children Elena and Elias, who tell me they're excited to read my book, even though I've assured them it's nowhere near as exciting as the *Harry Potter* and *Captain Underpants* books they're used to.

Dr. Wallis C. Metts Jr., my dad and the writer I admire most. My mother Katie, who has always encouraged me to pursue creativity and Christ. My brother Christian, who was the first person to show me how I could make a career out of words and websites.

Scott Kubie, who has shaped how I think about this work, and has been a constant, thoughtful friend.

My boss Susan Thome, who has taught me so much about leadership and collaboration.

Coworkers past and present who have helped me practice writing as design: Claire Rasmussen, Heather Ford-Helgeson, Brienne Moore, Johnny Taboada, Andrew Pulley, Jessica Zhang, Andrea Anibal, Julie Innes, Peter Shackelford, and Jeff Finley.

Andy Welfle, who has been an incredible coauthor and collaborator, helping me sharpen my perspective and showing me how to appreciate a good pencil. This couldn't have happened without you.

Andy would like to thank:

My ever-supportive partner, Katie Pruitt, for putting up with me for months, procrastinating and stealing myself away to work on the manuscript. And to Sebastian and Rupert, my cat-children, who give me a much-needed distraction when I get frustrated or stuck.

My parents, Lisa and Rick Welfle, and my sisters, Kelly Wade, Rosie Frayer, Nina Welfle, and Molly Welfle for their encouragement and love.

Dr. Johnny Gamber, Tim Wasem, Will Fanguy, and Harry Marks, my creative partners who let me ask them rapid-fire "research" questions.

My boss, Shawn Cheris, for taking a chance on me and letting me grow a practice while growing myself, and to my teammates at Adobe: Marisa Williams, Sarah Smart, Karissa Urry, Brandon Bussolini, Jess Sattell, Beth Anne Kinnaird, Davers Rosales, and Tessa Gregory for fighting the good UX writing fight and inspiring me with their work every day.

My former UX and content strategy managers, mentors, and colleagues: Nate Reusser, Rachel Gagnon, Erin Scime, Tony Headrick, Jonathon Colman, Kathy Matosich, Emily Shields and many, many more for their patience, guidance, and mentorship over the years. I wouldn't be able to do what I do if it wasn't for you.

And Michael Metts, for being such an inspiring and collaborative partner on this journey. There's literally no way I could have done it without you.

And we'd both like to thank:

Kristina Halvorson for introducing us to Lou Rosenfeld, our publisher. And to Lou and Marta from Rosenfeld Media, for taking a chance on us and helping us articulate and amplify our vision.

Devon Persing, John Caldwell, Sarah Smart, and Michael Haggerty-Villa, for the gift of their time and knowledge. Their interviews didn't make it into the book, but the perspective they shared was invaluable.

John Saito, Matt May, Ada Powers, Anna Pickard, Alicia Dougherty-Wold, Jasmine Probst, Lauren Lucchese, Katie Lower, Michaela Hackner, Melanie Polkosky, Hillary Accarizzi, Natalie Yee, and Jorge Arango for their time, insight, and generosity in making this a better book than if it was just our own voices.

Those who were kind enough to share their perspectives and career stories during our research for this book: Matthew Guay, Lindsey Phillips, Katherine Chimoy Vega, Niki Tores, and Greta Van der Merwe.

Our technical reviewers Andrea Drugay, Chelsea Larsson, Kathryn Strauss, Jonathon Colman, Rachel McConnell, Ryan Farrell, Scott Kubie, Susan Thome, and Sophie Tahran for their insight, feedback, and articulation. Their critique made this book so much better.

The UX and content strategy giants whose shoulders we stand on: Kristina Halvorson, Meghan Casey, Jonathon Colman, Sara Wachter-Boettcher, Erika Hall, Nicole Fenton, Kate Kiefer-Lee, Erin Scime, Sarah Richards, Erin Kissane, Karen McGrane, Steve Portigal, Jared Spool, Tiffani Jones Brown, and the many others who have helped us find our way in this field through their work.

Dear Reader,

Thanks very much for purchasing this book. There's a story behind it and every product we create at Rosenfeld Media.

Since the early 1990s, I've been a User Experience consultant, conference presenter, workshop instructor, and author. (I'm probably best-known for having cowritten *Information Architecture for the Web and Beyond*.) In each of these roles, I've been frustrated by the missed opportunities to apply UX principles and practices.

I started Rosenfeld Media in 2005 with the goal of publishing books whose design and development showed that a publisher could practice what it preached. Since then, we've expanded into producing industry-leading conferences and workshops. In all cases, UX has helped us create better, more successful products—just as you would expect. From employing user research to drive the design of our books and conference programs, to working closely with our conference speakers on their talks, to caring deeply about customer service, we practice what we preach every day.

Please visit **rosenfeldmedia.com** to learn more about our **conferences**, **workshops**, **free communities**, and **other great resources** that we've made for you. And send your ideas, suggestions, and concerns my way: louis@rosenfeldmedia.com

I'd love to hear from you, and I hope you enjoy the book!

Lou Rosenfeld,
Publisher

RECENT TITLES FROM ROSENFELD MEDIA

SELECTED TITLES FROM ROSENFELD MEDIA

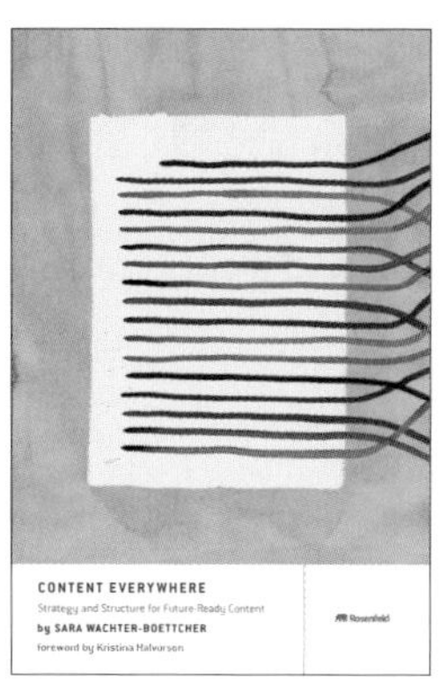

ABOUT THE AUTHORS

Michael J. Metts helps teams build great products and services by putting people first. With a background in journalism, he frequently finds himself talking about the role words play in designing useful, usable experiences. He has given talks and taught workshops on the topic at industry conferences around the world. He lives with his wife, two children, and a very small dog just outside Chicago.

When **Andy Welfle** was eight, he wanted to be a poet and a paleontologist. Twenty-eight years later, he is neither, but he uses those skills in his day job as a content strategist on Adobe's product design team—writing under huge constraints and uncovering artifacts from big, old software interfaces. When he's not working, he's creating podcasts and zines about one of his favorite topics: wooden pencils. Find him in San Francisco with his wife and two large cats.